Lecture Notes on Mathematical Olympiad Courses

For Junior Section Vol. 1

Mathematical Olympiad Series

ISSN: 1793-8570

Series Editors: Lee Peng Yee *(Nanyang Technological University, Singapore)*
Xiong Bin *(East China Normal University, China)*

Published

Vol. 1 A First Step to Mathematical Olympiad Problems
by Derek Holton (University of Otago, New Zealand)

Vol. 2 Problems of Number Theory in Mathematical Competitions
by Yu Hong-Bing (Suzhou University, China)
translated by Lin Lei (East China Normal University, China)

Vol. 3 Graph Theory
by Xiong Bin (East China Normal University, China) &
Zheng Zhongyi (High School Attached to Fudan University, China)
translated by Liu Ruifang, Zhai Mingqing & Lin Yuanqing
(East China Normal University, China)

Vol. 5 Selected Problems of the Vietnamese Olympiad (1962–2009)
by Le Hai Chau (Ministry of Education and Training, Vietnam)
& Le Hai Khoi (Nanyang Technology University, Singapore)

Vol. 6 Lecture Notes on Mathematical Olympiad Courses:
For Junior Section (In 2 Volumes)
by Xu Jiagu

Xu Jiagu

Vol. 6 Mathematical Olympiad Series

Lecture Notes on Mathematical Olympiad Courses

For Junior Section Vol. 1

Published by

World Scientific Publishing Co. Pte. Ltd.

5 Toh Tuck Link, Singapore 596224

USA office: 27 Warren Street, Suite 401-402, Hackensack, NJ 07601

UK office: 57 Shelton Street, Covent Garden, London WC2H 9HE

British Library Cataloguing-in-Publication Data
A catalogue record for this book is available from the British Library.

First published 2010
Reprinted with corrections 2011

Mathematical Olympiad Series — Vol. 6
LECTURE NOTES ON MATHEMATICAL OLYMPIAD COURSES
For Junior Section

Copyright © 2010 by World Scientific Publishing Co. Pte. Ltd.

All rights reserved. This book, or parts thereof, may not be reproduced in any form or by any means, electronic or mechanical, including photocopying, recording or any information storage and retrieval system now known or to be invented, without written permission from the Publisher.

For photocopying of material in this volume, please pay a copying fee through the Copyright Clearance Center, Inc., 222 Rosewood Drive, Danvers, MA 01923, USA. In this case permission to photocopy is not required from the publisher.

ISBN-13 978-981-4293-53-2 (pbk) (Set)
ISBN-10 981-4293-53-9 (pbk) (Set)

ISBN-13 978-981-4293-54-9 (pbk) (Vol. 1)
ISBN-10 981-4293-54-7 (pbk) (Vol. 1)

ISBN-13 978-981-4293-55-6 (pbk) (Vol. 2)
ISBN-10 981-4293-55-5 (pbk) (Vol. 2)

Printed in Singapore by World Scientific Printers.

Preface

Although mathematical olympiad competitions are carried out by solving problems, the system of Mathematical Olympiads and the related training courses cannot involve only the techniques of solving mathematical problems. Strictly speaking, it is a system of mathematical advancing education. To guide students who are interested in mathematics and have the potential to enter the world of Olympiad mathematics, so that their mathematical ability can be promoted efficiently and comprehensively, it is important to improve their mathematical thinking and technical ability in solving mathematical problems.

An excellent student should be able to think flexibly and rigorously. Here the ability to do formal logic reasoning is an important basic component. However, it is not the main one. Mathematical thinking also includes other key aspects, like starting from intuition and entering the essence of the subject, through prediction, induction, imagination, construction, design and their creative abilities. Moreover, the ability to convert concrete to the abstract and vice versa is necessary.

Technical ability in solving mathematical problems does not only involve producing accurate and skilled-computations and proofs, the standard methods available, but also the more unconventional, creative techniques.

It is clear that the usual syllabus in mathematical educations cannot satisfy the above requirements, hence the mathematical olympiad training books must be self-contained basically.

The book is based on the lecture notes used by the editor in the last 15 years for Olympiad training courses in several schools in Singapore, like Victoria Junior college, Hwa Chong Institution, Nanyang Girls High School and Dunman High School. Its scope and depth significantly exceeds that of the usual syllabus, and introduces many concepts and methods of modern mathematics.

The core of each lecture are the concepts, theories and methods of solving mathematical problems. Examples are then used to explain and enrich the lectures, and indicate their applications. And from that, a number of questions are included for the reader to try. Detailed solutions are provided in the book.

The examples given are not very complicated so that the readers can understand them more easily. However, the practice questions include many from actual competitions which students can use to test themselves. These are taken from a range of countries, e.g. China, Russia, the USA and Singapore. In particular, there are many questions from China for those who wish to better understand mathematical Olympiads there. The questions are divided into two parts. Those in Part A are for students to practise, while those in Part B test students' ability to apply their knowledge in solving real competition questions.

Each volume can be used for training courses of several weeks with a few hours per week. The test questions are not considered part of the lectures, since students can complete them on their own.

Acknowledgments

My thanks to Professor Lee Peng Yee for suggesting the publication of this the book and to Professor Phua Kok Khoo for his strong support. I would also like to thank my friends, Ang Lai Chiang, Rong Yifei and Gwee Hwee Ngee, lecturers at HwaChong, Tan Chik Leng at NYGH, and Zhang Ji, the editor at WSPC for her careful reading of my manuscript, and their helpful suggestions. This book would be not published today without their efficient assistance.

Abbreviations and Notations

Abbreviations

AHSME	American High School Mathematics Examination
AIME	American Invitational Mathematics Examination
APMO	Asia Pacific Mathematics Olympiad
ASUMO	Olympics Mathematical Competitions of All the Soviet Union
AUSTRALIA	Australia Mathematical Competitions
BMO	British Mathematical Olympiad
CHNMO	China Mathematical Olympiad
CHNMOL	China Mathematical Competition for Secondary Schools
CHINA	China Mathematical Competitions for Secondary Schools except for CHNMOL
CMO	Canada Mathematical Olympiad
HUNGARY	Hungary Mathematical Competition
IMO	International Mathematical Olympiad
JAPAN	Japan Mathematical Olympiad
KIEV	Kiev Mathematical Olympiad
MOSCOW	Moscow Mathematical Olympiad
NORTH EUROPE	North Europe Mathematical Olympiad
RUSMO	All-Russia Olympics Mathematical Competitions
SSSMO	Singapore Secondary Schools Mathematical Olympiads
SMO	Singapore Mathematical Olympiads
SSSMO(J)	Singapore Secondary Schools Mathematical Olympiads for Junior Section
UKJMO	United Kingdom Junior Mathematical Olympiad
USAMO	United States of American Mathematical Olympiad

Notations for Numbers, Sets and Logic Relations

$\mathbb{N}$	the set of positive integers (natural numbers)
$\mathbb{N}_0$	the set of non-negative integers
$\mathbb{Z}$	the set of integers
$\mathbb{Z}^+$	the set of positive integers
$\mathbb{Q}$	the set of rational numbers
$\mathbb{Q}^+$	the set of positive rational numbers
$\mathbb{Q}_0^+$	the set of non-negative rational numbers
$\mathbb{R}$	the set of real numbers
$[a, b]$	the closed interval, i.e. all x such that $a \leq x \leq b$
(a, b)	the open interval, i.e. all x such that $a < x < b$
$\Leftrightarrow$	iff, if and only if
$\Rightarrow$	implies
$A \subset B$	A is a subset of B
$A - B$	the set formed by all the elements in A but not in B
$A \cup B$	the union of the sets A and B
$A \cap B$	the intersection of the sets A and B
$a \in A$	the element a belongs to the set A

Contents

Lecture 1

Operations on Rational Numbers

1. **Basic Rules on Addition, Subtraction, Multiplication, Division**
   Commutative Law: $a + b = b + a \quad ab = ba$
   Associative Law: $a + b + c = a + (b + c) \quad (ab)c = a(bc)$
   Distributive Law: $ac + bc = (a + b)c = c(a + b)$

2. **Rule for Removing Brackets**
   For any rational numbers x, y,
   (i) $x + (y) = x + y, \quad x + (-y) = x - y;$
   (ii) $x - (y) = x - y, x - (-y) = x + y.$
   (iii) $x \times (-y) = -xy; \quad (-x) \times y = -xy; \quad (-x) \times (-y) = xy;$
   $(-1)^n = -1$ for odd n, $(-1)^n = 1$ for even n.
   (iv) If the denominators of the following expressions are all not zeros, then
   $$\frac{x}{-y} = -\frac{x}{y}; \quad \frac{-x}{y} = -\frac{x}{y}; \quad \frac{-x}{-y} = \frac{x}{y}.$$

3. **Ingenious Ways for Calculating**
   - Make a **telescopic sum** by using the following expressions:
   $$\frac{1}{k(k+1)} = \frac{1}{k} - \frac{1}{k+1},$$
   $$\frac{1}{k(k+m)} = \frac{1}{m}\left(\frac{1}{k} - \frac{1}{k+m}\right),$$
   $$\frac{1}{k(k+1)(k+2)} = \frac{1}{2}\left[\frac{1}{k(k+1)} - \frac{1}{(k+1)(k+2)}\right].$$
   - By use of the following formulae:
   $$(a \pm b)^2 = a^2 + 2ab + b^2;$$
   $$a^2 - b^2 = (a - b)(a + b);$$
   $$a^3 + b^3 = (a + b)(a^2 - ab + b^2);$$
   $$a^3 - b^3 = (a - b)(a^2 + ab + b^2), \text{etc.}$$

Examples

Example 1. Evaluate $(-5)^2 \times \left(-\frac{1}{5}\right)^3 - 2^3 \div \left(-\frac{1}{2}\right)^2 - (-1)^{1999}$.

Solution $(-5)^2 \times \left(-\frac{1}{5}\right)^3 - 2^3 \div \left(-\frac{1}{2}\right)^2 - (-1)^{1999}$

$$= 5^2 \times \left(-\frac{1}{125}\right) - 8 \div \frac{1}{4} - (-1)$$

$$= -\frac{1}{5} - 8 \times 4 + 1 = -\frac{1}{5} - 31 = -31\frac{1}{5}.$$

Example 2. There are five operational expressions below:

(i) $(2 \times 3 \times 5 \times 7)\left(\frac{1}{2} + \frac{1}{3} + \frac{1}{5} + \frac{1}{7}\right)$;

(ii) $(-0.125)^7 \cdot 8^8$;

(iii) $(-11) + (-33) - (-55) - (-66) - (-77) - (-88)$;

(iv) $\left(-\frac{75}{13}\right)^2 + \left(\frac{37}{13}\right)^2$;

(v) $\left[\left(-\frac{6}{7}\right)^7 + \left(-\frac{4}{5}\right) \times \left(-\frac{4}{9}\right) \times \frac{16}{81}\right] \times \left(9\frac{246}{247} - 0.666\right)$.

Then the expression with maximal value is

(A) (i), (B) (iii), (C) (iv), (D) (v).

Solution

(i) $(2 \times 3 \times 5 \times 7)\left(\frac{1}{2} + \frac{1}{3} + \frac{1}{5} + \frac{1}{7}\right)$

$= 105 + 70 + 42 + 30 = 247$;

(ii) $(-0.125)^7 \cdot 8^8 = -(0.125 \times 8)^7 \times 8 = -8$;

(iii) $(-11) + (-33) - (-55) - (-66) - (-77) - (-88)$

$= -11 - 33 + 55 + 66 + 77 + 88 = 11 \times 22 = 242$;

(iv) $\left(-\frac{75}{13}\right)^2 + \left(\frac{37}{13}\right)^2 < 6^2 + 3^2 = 45$;

(v) $\left[\left(-\frac{6}{7}\right)^7 + \left(-\frac{4}{5}\right) \times \left(-\frac{4}{9}\right) \times \frac{16}{81}\right] \times \left(9\frac{246}{247} - 0.666\right)$

$< 1 \times 10 = 10$;

Thus, the answer is (A).

Example 3. $123456789 \times 999999999 =$ ________.

Solution

$$123456789 \times 999999999 = 123456789 \times (1000000000 - 1)$$
$$= 123456789000000000 - 123456789 = 123456788876543211.$$

Example 4. The value of $\dfrac{13579}{(-13579)^2 + (-13578)(13580)}$ is
(A) 1, (B) 13579, (C) -1, (D) -13578.

Solution By use of $(a - b)(a + b) = a^2 - b^2$, we have

$$\frac{13579}{(-13579)^2 + (-13578)(13580)} = \frac{13579}{(13579)^2 - (13579^2 - 1)} = 13579.$$

The answer is (B).

Example 5. $\dfrac{83^3 + 17^3}{83 \times 66 + 17^2} =$ ________.

Solution By use of the formula $a^3 + b^3 = (a + b)(a^2 - ab + b^2)$,

$$\frac{83^3 + 17^3}{83 \times 66 + 17^2} = \frac{(83 + 17)(83^2 - 83 \times 17 + 17^2)}{83 \times 66 + 17^2} = \frac{100 \times (83 \times 66 + 17^2)}{83 \times 66 + 17^2} = 100.$$

Example 6. Evaluate

$$\frac{(4 \times 7 + 2)(6 \times 9 + 2)(8 \times 11 + 2) \cdot \cdots \cdot (100 \times 103 + 2)}{(5 \times 8 + 2)(7 \times 10 + 2)(9 \times 12 + 2) \cdot \cdots \cdot (99 \times 102 + 2)}.$$

Solution From $n(n + 3) + 2 = n^2 + 3n + 2 = (n + 1)(n + 2)$ for any integer n, we have

$$\frac{(4 \times 7 + 2)(6 \times 9 + 2)(8 \times 11 + 2) \cdot \cdots \cdot (100 \times 103 + 2)}{(5 \times 8 + 2)(7 \times 10 + 2)(9 \times 12 + 2) \cdot \cdots \cdot (99 \times 102 + 2)}$$
$$= \frac{(5 \times 6)(7 \times 8)(9 \times 10) \cdot \cdots \cdot (101 \times 102)}{(6 \times 7)(8 \times 9)(10 \times 11) \cdot \cdots \cdot (100 \times 101)}$$
$$= 5 \times 102 = 510.$$

Example 7. $\dfrac{20092008^2}{20092007^2 + 20092009^2 - 2} =$ ________.

Solution $\dfrac{20092008^2}{20092007^2+20092009^2-2}$

$$= \frac{20092008^2}{(20092007^2-1)+(20092009^2-1)}$$
$$= \frac{20092008^2}{(20092006)(20092008)+(20092008)(20092010)}$$
$$= \frac{20092008^2}{(20092008)(20092006+20092010)} = \frac{20092008^2}{2(20092008^2)} = \frac{1}{2}.$$

Example 8. $3-\frac{1}{2}-\frac{1}{6}-\frac{1}{12}-\frac{1}{20}-\frac{1}{30}-\frac{1}{42}-\frac{1}{56}=$ ________.

Solution

$$3-\left(\frac{1}{2}+\frac{1}{6}+\frac{1}{12}+\frac{1}{20}+\frac{1}{30}+\frac{1}{42}+\frac{1}{56}\right)$$
$$=3-\left(\frac{1}{1\times 2}+\frac{1}{2\times 3}+\frac{1}{3\times 4}+\cdots+\frac{1}{7\times 8}\right)$$
$$=3-\left[\left(1-\frac{1}{2}\right)+\left(\frac{1}{2}-\frac{1}{3}\right)+\cdots+\left(\frac{1}{7}-\frac{1}{8}\right)\right]$$
$$=3-\left(1-\frac{1}{8}\right)=2\frac{1}{8}.$$

Example 9. Evaluate $\frac{1}{3}+\frac{1}{15}+\frac{1}{35}+\frac{1}{63}+\frac{1}{99}+\frac{1}{143}$.

Solution Since $\dfrac{1}{k(k+2)}=\dfrac{1}{2}\left(\dfrac{1}{k}-\dfrac{1}{k+2}\right)$ for any positive integer k, so

$$\frac{1}{3}+\frac{1}{15}+\frac{1}{35}+\frac{1}{63}+\frac{1}{99}+\frac{1}{143}$$
$$=\frac{1}{1\times 3}+\frac{1}{3\times 5}+\frac{1}{5\times 7}+\frac{1}{7\times 9}+\frac{1}{9\times 11}+\frac{1}{11\times 13}$$
$$=\frac{1}{2}\left[\left(\frac{1}{1}-\frac{1}{3}\right)+\left(\frac{1}{3}-\frac{1}{5}\right)+\cdots+\left(\frac{1}{11}-\frac{1}{13}\right)\right]$$
$$=\frac{1}{2}\times\left[1-\frac{1}{13}\right]=\frac{6}{13}.$$

Example 10. If $ab<0$, then the relation in sizes of $(a-b)^2$ and $(a+b)^2$ is
(A) $(a-b)^2<(a+b)^2$; (B) $(a-b)^2=(a+b)^2$;
(C) $(a-b)^2>(a+b)^2$; (D) not determined.

Solution From $(a-b)^2 = a^2 - 2ab + b^2 = a^2 + 2ab + b^2 - 4ab = (a+b)^2 - 4ab > (a+b)^2$, the answer is (C).

Example 11. If $-1 < a < 0$, then the relation in sizes of $a^3, -a^3, a^4, -a^4, \frac{1}{a}, -\frac{1}{a}$ is

(A) $\frac{1}{a} < -a^4 < a^3 < -a^3 < a^4 < -\frac{1}{a}$;

(B) $a < \frac{1}{a} < -a^4 < a^4 < -\frac{1}{a} < -a^3$;

(C) $\frac{1}{a} < a^3 < -a^4 < a^4 < -a^3 < -\frac{1}{a}$;

(D) $\frac{1}{a} < a^3 < a^4 < -a^4 < -a^3 < -\frac{1}{a}$.

Solution From $-1 < a < 0$ we have $0 < a^4 < -a^3 < 1 < -\frac{1}{a}$, so $-a^4 > a^3$ and $-\frac{1}{a} > -a^3$ and $a^4 > -a^4$, the answer is (C).

Testing Questions (A)

1. Evaluate $-1 - (-1)^1 - (-1)^2 - (-1)^3 - \cdots - (-1)^{99} - (-1)^{100}$.

2. Evaluate $2008 \times 20092009 - 2009 \times 20082008$.

3. From 2009 subtract half of it at first, then subtract $\frac{1}{3}$ of the remaining number, next subtract $\frac{1}{4}$ of the remaining number, and so on, until $\frac{1}{2009}$ of the remaining number is subtracted. What is the final remaining number?

4. Find the sum $\frac{1}{5\times 7} + \frac{1}{7\times 9} + \frac{1}{9\times 11} + \frac{1}{11\times 13} + \frac{1}{13\times 15}$.

5. Find the sum $\frac{1}{10} + \frac{1}{40} + \frac{1}{88} + \frac{1}{154} + \frac{1}{238}$.

6. Evaluate

$$\left(\frac{1}{3} + \frac{1}{4} + \cdots + \frac{1}{2009}\right)\left(1 + \frac{1}{2} + \cdots + \frac{1}{2008}\right) - \left(1 + \frac{1}{3} + \frac{1}{4} + \cdots + \frac{1}{2009}\right)\left(\frac{1}{2} + \frac{1}{3} + \cdots + \frac{1}{2008}\right).$$

7. Find the sum $\dfrac{1}{1+2}+\dfrac{1}{1+2+3}+\cdots+\dfrac{1}{1+2+\cdots+51}$.

8. Let n be a positive integer, find the value of

$$1+\frac{1}{2}+\frac{2}{2}+\frac{1}{2}+\frac{1}{3}+\frac{2}{3}+\frac{3}{3}+\frac{2}{3}+\frac{1}{3}+\cdots+\frac{1}{n}+\frac{2}{n}+\cdots+\frac{n}{n}+\frac{n-1}{n}+\cdots+\frac{1}{n}.$$

9. Evaluate $1^2-2^2+3^2-4^2+\cdots-2008^2+2009^2$.

10. Find the sum $11+192+1993+19994+199995+1999996+19999997+199999998+1999999999$.

Testing Questions (B)

1. Calculate $\dfrac{3^2+1}{3^2-1}+\dfrac{5^2+1}{5^2-1}+\dfrac{7^2+1}{7^2-1}+\cdots+\dfrac{99^2+1}{99^2-1}$.

2. After simplification, the value of

$$1-\frac{2}{1\cdot(1+2)}-\frac{3}{(1+2)(1+2+3)}-\frac{4}{(1+2+3)(1+2+3+4)}$$
$$-\cdots-\frac{100}{(1+2+\cdots+99)(1+2+\cdots+100)}$$

is a proper fraction in its lowest form. Find the difference of its denominator and numerator.

3. Evaluate $\dfrac{1}{1\times2\times3}+\dfrac{1}{2\times3\times4}+\cdots+\dfrac{1}{100\times101\times102}$.

4. Find the sum

$$\frac{1}{1+1^2+1^4}+\frac{2}{1+2^2+2^4}+\frac{3}{1+3^2+3^4}+\cdots+\frac{50}{1+50^2+50^4}.$$

5. Evaluate the expression

$$\frac{1^2}{1^2-10+50}+\frac{2^2}{2^2-20+50}+\cdots+\frac{9^2}{9^2-90+50}.$$

Lecture 2

Monomials and Polynomials

Definitions

Monomial: A product of numerical numbers and letters is said to be a monomial. In particular, a number or a letter alone is also a monomial, for example, $16, 32x$, and $2ax^2y$, etc.

Coefficient: In each monomial, the part consisting of numerical numbers and the letters denoting constants is said to be the coefficient of the monomial, like 32 in $32x$, $2a$ in $2ax^2y$, etc.

Degree of a Monomial: In a monomial, the sum of all indices of the letters denoting variables is called the degree of the monomial. For example, the degree of $3abx^2$ is 2, and the degree of $7a^4xy^2$ is 3.

Polynomial: The sum of several monomials is said to be a polynomial, its each monomial is called a **term**, the term not containing letters is said to be the **constant term** of the polynomial. The maximum value of the degree of terms in the polynomial is called **degree of the polynomial**, for example, the degree is 2 for $3x^2 + 4x + 1$, and 5 for $2x^2y^3 + 2y$. A polynomial is called **homogeneous** when all its terms have the same degree, like $3x^2 + xy + 4y^2$.

Arrangement of Terms: When arranging the terms in a polynomial, the terms can be arranged such that their degrees are in either ascending or descending order, and the sign before a term should remain attached to when moving it. For example, the polynomial $x^3y^3 - 1 - 2xy^2 - x^3y$ should be arranged as $x^3y^3 - x^3y - 2xy^2 - 1$ or $-1 - 2xy^2 - x^3y + x^3y^3$.

Like Terms: Two terms are called like terms if they have the same construction except for their coefficients, like in $4ax^2y$ and $5bx^2y$.

Combining Like Terms: When doing addition, subtraction to two like terms, it means doing the corresponding operation on their coefficients. For example, $4ax^2y + 5bx^2y = (4a + 5b)x^2y$ and $4ax^2y - 5bx^2y = (4a - 5b)x^2y$.

Operations on Polynomials

Addition: Adding two polynomials means:
(i) take all terms in the two polynomials as the terms of the sum;
(ii) combine all the like terms if any;
(iii) arrange all the combined terms according to the order of ascending or descending degree.

Subtraction: Let P and Q be two polynomials. Then $P - Q$ means
(i) change the signs of all terms in Q to get $-Q$ at first;
(ii) take all terms in the two polynomials P and $-Q$ as the terms of $P - Q$;
(iii) combine all the like terms if any;
(iv) arrange all the combined terms according to the rule mentioned above.

Rule for Removing or Adding Brackets:
The rule for removing or adding brackets is the distributive law. For example, to remove the brackets in the expression $-2x(x^3y - 4x^2y^2 + 4)$, then

$$-2x(x^3y - 4x^2y^2 + 4) = -2x^4y + 8x^3y^2 - 8x,$$

and to add a pair of bracket for containing the terms of the expression $-4x^5y^2 + 6x^4y - 8x^2y^2$ and pick out their common factor with negative coefficient, then

$$-4x^5y^2 + 6x^4y - 8x^2y^2 = -2x^2y(2x^3y - 3x^2 + 4y).$$

Multiplication:
(i) For natural numbers m and n,
$a^m \cdot a^n = a^{m+n}$; $(a^m)^n = a^{mn}$; $(ab)^n = a^n b^n$;
(ii) When two monomials are multiplied, the coefficient of the product is the product of the coefficients, the letters are multiplied according to the rules in (i);
(iii) When two polynomials are multiplied, by using the distributive law, get a sum of products of a monomial and a polynomial first, and then use the distributive law again, get a sum of products of two monomials;
(iv) Three basic formulae in multiplication:
(i) $(a - b)(a + b) = a^2 - b^2$;
(ii) $(a + b)^2 = a^2 + 2ab + b^2$;
(iii) $(a - b)^2 = a^2 - 2ab + b^2$.

Examples

Example 1. Simplify $3a + \{-4b - [4a - 7b - (-4a - b)] + 5a\}$.

Solution

$$3a + \{-4b - [4a - 7b - (-4a - b)] + 5a\}$$
$$= 3a + \{-4b - [8a - 6b] + 5a\} = 3a + \{-3a + 2b\} = 2b.$$

or

$$3a + \{-4b - [4a - 7b - (-4a - b)] + 5a\}$$
$$= 8a - 4b - [4a - 7b - (-4a - b)] = 4a + 3b + (-4a - b) = 2b.$$

Note: We can remove the brackets from the innermost to outermost layer, or *vice versa*.

Example 2. Simplify the expression $4\{(3x-2) - [3(3x-2)+3]\} - (4-6x)$.

Solution Taking $3x - 2$ as whole as one number y in the process of the simplification first, we have

$$4\{(3x-2) - [3(3x-2)+3]\} - (4-6x) = 4\{y - [3y+3]\} + 2y$$
$$= 4\{-2y - 3\} + 2y = -6y - 12 = -6(3x-2) - 12 = -18x.$$

Example 3. Evaluate $-9x^{n-2} - 8x^{n-1} - (-9x^{n-2}) - 8(x^{n-2} - 2x^{n-1})$, where $x = 9,\ n = 3$.

Solution $-9x^{n-2} - 8x^{n-1} - (-9x^{n-2}) - 8(x^{n-2} - 2x^{n-1}) = 8x^{n-1} - 8x^{n-2}$. By substituting $x = 9, n = 3$, it follows that

$$\text{the expression} = 8x^{n-1} - 8x^{n-2} = 8 \times (81 - 9) = 576.$$

Example 4. Given $x^3 + 4x^2y + axy^2 + 3xy - bx^cy + 7xy^2 + dxy + y^2 = x^3 + y^2$ for any real numbers x and y, find the value of a, b, c, d.

Solution $4x^2y$ and $-bx^cy$ must be like terms and their sum is 0, so

$$b = 4, c = 2.$$

$axy^2 + 7xy^2 = 0$ and $2xy + dxy = 0$ for every x and y yields $a + 7 = 0$ and $3 + d = 0$, so

$$a = -7, d = -3.$$

Thus, $a = -7, b = 4, c = 2, d = -3$.

Example 5. Given that m, x, y satisfy (i) $\frac{2}{3}(x-5)^2 + 5m^2 = 0$; (ii) $-2a^2b^{y+1}$ and $3a^2b^3$ are like terms, find the value of the expression

$$\frac{3}{8}x^2y + 5m^2 - \left\{-\frac{7}{16}x^2y + \left[-\frac{1}{4}xy^2 - \frac{3}{16}x^2y - 3.475xy^2\right] - 6.275xy^2\right\}.$$

Solution The condition (i) implies $(x-5)^2 = 0, 5m^2 = 0$, so $x = 5, m = 0$. The condition (ii) implies $y + 1 = 3$, i.e. $y = 2$. Therefore

$$\begin{aligned}
&\frac{3}{8}x^2y + 5m^2 - \left\{-\frac{7}{16}x^2y + \left[-\frac{1}{4}xy^2 - \frac{3}{16}x^2y - 3.475xy^2\right] - 6.275xy^2\right\} \\
&= \frac{3}{8}x^2y - \left\{-\frac{7}{16}x^2y - \frac{1}{4}xy^2 - \frac{3}{16}x^2y - 3.475xy^2 - 6.275xy^2\right\} \\
&= \frac{3}{8}x^2y + \frac{7}{16}x^2y + \frac{1}{4}xy^2 + \frac{3}{16}x^2y + 3.475xy^2 + 6.275xy^2 \\
&= \left(\frac{3}{8} + \frac{7}{16} + \frac{3}{16}\right)x^2y + \left(\frac{1}{4} + 3\frac{19}{40} + 6\frac{11}{40}\right)xy^2 \\
&= x^2y + 10xy^2 = (5^2)(2) + 10(5)(2^2) = 250.
\end{aligned}$$

Example 6. Given that $P(x) = nx^{n+4} + 3x^{4-n} - 2x^3 + 4x - 5$, $Q(x) = 3x^{n+4} - x^4 + x^3 + 2nx^2 + x - 2$ are two polynomials. Determine if there exists an integer n such that the difference $P - Q$ is a polynomial with degree 5 and six terms.

Solution $P(x) - Q(x) = (n-3)x^{n+4} + 3x^{4-n} + x^4 - 3x^3 - 2nx^2 + 3x - 3$.

When $n + 4 = 5$, then $n = 1$, so that $3x^{4-n} - 3x^3 = 0$, the difference has only 5 terms.

When $4 - n = 5$, then $n = -1$, so that $P(x) - Q(x) = 3x^5 + x^4 - 7x^3 + 2x^2 + 3x - 3$ which satisfies the requirement. Thus, $n = -1$.

Example 7. Expand $(x-1)(x-2)(x-3)(x-4)$.

Solution

$$\begin{aligned}
&(x-1)(x-2)(x-3)(x-4) = [(x-1)(x-4)] \cdot [(x-2)(x-3)] \\
&= (x^2 - 4x - x + 4)(x^2 - 3x - 2x + 6) \\
&= [(x^2 - 5x + 5) - 1][(x^2 - 5x + 5) + 1] \\
&= (x^2 - 5x + 5)^2 - 1 = x^4 + 25x^2 + 25 - 10x^3 + 10x^2 - 50x - 1 \\
&= x^4 - 10x^3 + 35x^2 - 50x + 24.
\end{aligned}$$

Example 8. Expand $\left(5xy - 3x^2 + \frac{1}{2}y^2\right)\left(5xy + 3x^2 - \frac{1}{2}y^2\right)$

Solution Considering the formula $(a-b)(a+b) = a^2 - b^2$, we have

$$\begin{aligned}
&\left(5xy - 3x^2 + \frac{1}{2}y^2\right)\left(5xy + 3x^2 - \frac{1}{2}y^2\right) \\
&= \left[5xy - \left(3x^2 - \frac{1}{2}y^2\right)\right] \cdot \left[5xy + \left(3x^2 - \frac{1}{2}y^2\right)\right]
\end{aligned}$$

$$= 25x^2y^2 - \left(3x^2 - \frac{1}{2}y^2\right)^2 = 25x^2y^2 - \left[9x^4 - 3x^2y^2 + \left(\frac{1}{2}y^2\right)^2\right]$$

$$= 25x^2y^2 - \left[9x^4 - 3x^2y^2 + \frac{1}{4}y^4\right] = -9x^4 + 28x^2y^2 - \frac{1}{4}y^2.$$

Example 9. Given $x^2 - x - 1 = 0$, simplify $\dfrac{x^3 + x + 1}{x^5}$ to a polynomial form.

Solution $x^2 - x - 1 = 0$ yields $x + 1 = x^2$, so

$$\frac{x^3 + x + 1}{x^5} = \frac{x^3 + x^2}{x^5} = \frac{x + 1}{x^3} = \frac{1}{x} = \frac{x^2 - x}{x} = x - 1.$$

Testing Questions (A)

1. In the following expressions, which is (are) not monomial?

   (A) $\dfrac{x}{5}$ (B) $-0.5(1 + \dfrac{1}{x})$ (C) $\dfrac{3}{x^2}$

2. The degree of sum of two polynomials with degree 4 each must be

   (A) 8, (B) 4, (C) less than 4, (D) not greater than 4.

3. While doing an addition of two polynomials, Adam mistook "add the polynomial $2x^2 + x + 1$" as "subtract $2x^2 + x + 1$", and hence his result was $5x^2 - 2x + 4$. Find the correct answer.

4. Given that the monomials $0.75x^by^c$ and $-0.5x^{m-1}y^{2n-1}$ are like terms, and their sum is $1.25ax^ny^m$, find the value of abc.

5. If $x^5, x + \dfrac{1}{x}, 1 + \dfrac{2}{x} + \dfrac{3}{x^2}$ are multiplied together, the product is a polynomial, then degree of the product is

   (A) 4, (B) 5, (C) 6, (D) 7, (E) 8.

6. Find a natural number n, such that $2^8 + 2^{10} + 2^n$ is a perfect square number.

7. Given $3x^2 + x = 1$, find the value of $6x^3 - x^2 - 3x + 2010$.

8. If $x = \dfrac{a}{b+c} = \dfrac{b}{a+c} = \dfrac{c}{a+b}$, then the value of x is

   (A) $\dfrac{1}{2}$, (B) -1, (C) $\dfrac{1}{2}$, or -1, (D) $\dfrac{3}{2}$.

9. If $\frac{1}{x} - \frac{1}{y} = 4$, find the value of $\frac{2x + 4xy - 2y}{x - y - 2xy}$.

Testing Questions (B)

1. (UKJMO/1995(B)) Nine squares are arranged to form a rectangle as shown. The smallest square has side of length 1. How big is the next smallest square? and how about the area of the rectangle?

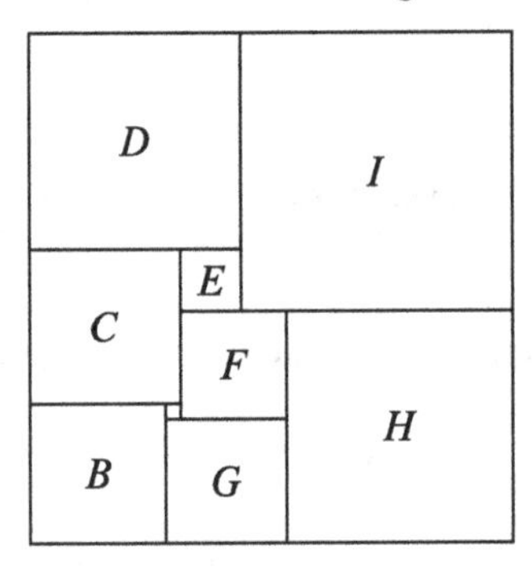

2. Let $P(x) = ax^7 + bx^3 + cx - 5$, where a, b, c are constants. Given $P(-7) = 7$, find the value of $P(7)$.

3. If a, b, c are non-zero real numbers, satisfying $\frac{1}{a} + \frac{1}{b} + \frac{1}{c} = \frac{1}{a+b+c}$, prove that among a, b, c there must be two opposite numbers.

4. If $xy = a, xz = b, yz = c$ and $abc \neq 0$, find the value of $x^2 + y^2 + z^2$ in terms of a, b, c.

5. Given $a^4 + a^3 + a^2 + a + 1 = 0$. Find the value of $a^{2000} + a^{2010} + 1$.

6. If $(x^2 - x - 1)^n = a_{2n}x^{2n} + a_{2n-1}x^{2n-1} + \cdots + a_2x^2 + a_1x + a_0$, find the value of $a_0 + a_2 + a_4 + \cdots + a_{2n}$.

Lecture 3

Linear Equations of Single Variable

Usual Steps for Solving Equations

(i) **Remove denominators**: When each term of the given equation is multiplied by the L.C.M. of denominators, all the denominators of the terms can be removed. After removing the denominators, the numerator of each term is considered as whole as an algebraic expression, and should be put in brackets.

(ii) **Remove brackets**: We can remove brackets by using the distributive law and the rules for removing brackets. Do not leave out any term inside the brackets, and change the signs of **each** term inside the brackets if there is a "−" sign before the brackets.

(iii) **Move terms**: Move all the terms with unknown variable to one side of the equation and other terms to another side of the equation according to the **Principle for moving terms**: when moving a term from one side to the other side of an equation, its sign must be changed. All unmoved terms keep their signs unchanged.

(iv) **Combine like terms**: After moving the terms, the like terms should be combined, so that the given equation is in the form

$$ax = b$$

where a, b are constants but sometimes are unknown. An unknown constant in an equation is called a **parameter**

(v) **Normalize the coefficient of x**: When $a \neq 0$, we have unique solution $x = \frac{b}{a}$. If $a = 0$ but $b \neq 0$, the equation has no solution. If $a = b = 0$, any real value is a solution for x. In particular, when a contains parameters, a cannot be moved to the right as denominator unless it is not zero, and thus it is needed to discuss the value of a on a case by case basis.

Remark: It is not needed to perform the above steps according to the order listed strictly, different orders are needed for different questions.

Examples

Example 1. Solve the equation $\frac{1}{10}\left\{\frac{1}{9}\left[\frac{1}{5}\left(\frac{x+2}{3}+8\right)+16\right]+8\right\}=1.$

Solution By removing the denominators one-by-one, it follows that

$$\frac{1}{9}\left[\frac{1}{5}\left(\frac{x+2}{3}+8\right)+16\right]=2,$$

$$\frac{1}{5}\left(\frac{x+2}{3}+8\right)=2$$

$$\frac{x+2}{3}=2,$$

$$\therefore x+2=6, \text{ i.e. } x=4.$$

Example 2. Solve the equation $\frac{1}{5}\left\{\frac{1}{4}\left[\frac{1}{3}\left(\frac{1}{2}x-3\right)-2\right]-1\right\}-2=1.$

Solution Here it's convenient to move a term and then remove a denominator for simplifying the equation. From the given equation we have

$$\frac{1}{5}\left\{\frac{1}{4}\left[\frac{1}{3}\left(\frac{1}{2}x-3\right)-2\right]-1\right\}-2=1,$$

$$\frac{1}{4}\left[\frac{1}{3}\left(\frac{1}{2}x-3\right)-2\right]-1=15,$$

$$\frac{1}{3}\left(\frac{1}{2}x-3\right)-2=64$$

$$\frac{1}{2}x-3=198$$

$$\therefore x=402.$$

Example 3. Solve the equation $\frac{3}{5}\left[\frac{5}{3}\left(\frac{1}{4}x+1\right)+5\right]-\frac{1}{2}=x.$

Solution Considering that $\frac{5}{3}$ and $\frac{3}{5}$ are reciprocal each other, it is better to remove the brackets first. We have

$$\left(\frac{1}{4}x+1\right)+3-\frac{1}{2}=x,$$

$$\frac{1}{4}x+1+3-\frac{1}{2}=x,$$

$$\frac{3}{4}x=\frac{7}{2}, \quad \therefore x=\frac{14}{3}.$$

Example 4. Solve the equation $1-\dfrac{x-\dfrac{1+3x}{5}}{3}=\dfrac{x}{2}-\dfrac{2x-\dfrac{10-6x}{7}}{2}$.

Solution Since the given equation contains complex fractions in both sides, it is better to simplify each side separately first. From

$$1-\frac{x-\frac{1+3x}{5}}{3}=1-\frac{5x-(1+3x)}{15}=\frac{15-2x+1}{15}=\frac{16-2x}{15},$$
$$\frac{x}{2}-\frac{2x-\frac{10-6x}{7}}{2}=\frac{x}{2}-\frac{14x-(10-6x)}{14}=\frac{10-13x}{14},$$

it follows that

$$\frac{16-2x}{15}=\frac{10-13x}{14},$$
$$14(16-2x)=15(10-13x),$$
$$224-28x=150-195x,\ \text{ i.e. }\ 167x=-74,$$
$$\therefore x=-\frac{74}{167}.$$

Example 5. If a, b, c are positive constants, solve the equation

$$\frac{x-a-b}{c}+\frac{x-b-c}{a}+\frac{x-c-a}{b}=3.$$

Solution By moving 3 in the given equation to the left hand side, it follows that

$$\left(\frac{x-a-b}{c}-1\right)+\left(\frac{x-b-c}{a}-1\right)+\left(\frac{x-c-a}{b}-1\right)=0,$$
$$\frac{x-a-b-c}{c}+\frac{x-a-b-c}{a}+\frac{x-a-b-c}{b}=0,$$
$$(x-a-b-c)\left(\frac{1}{c}+\frac{1}{a}+\frac{1}{b}\right)=0,$$
$$\because \frac{1}{c}+\frac{1}{a}+\frac{1}{b}>0, \therefore x-a-b-c=0,\ \text{ i.e. }\ x=a+b+c.$$

Example 6. Solve the equation $ax+b-\dfrac{5x+2ab}{5}=\dfrac{1}{4}$.

Solution Removing the denominators of the given equation yields

$$20(ax+b)-4(5x+2ab)=5,$$
$$20ax+20b-20x-8ab=5,$$
$$20(a-1)x=5-20b+8ab.$$

(i) When $a \neq 1$, $x = \dfrac{5-20b+8ab}{20(a-1)}$.

(ii) When $a = 1$ and $b = \dfrac{5}{12}$, the equation becomes $0 \cdot x = 0$, so any real number is a solution for x.

(iii) When $a = 1$ and $b \neq \dfrac{5}{12}$, the equation becomes $0 \cdot x = 5 - 12b$, so no solution for x.

Example 7. Given that the equation $a(2x+3)+3bx = 12x+5$ has infinitely many solutions for x. Find the values of a and b.

Solution Change the given equation to the form $(2a+3b-12)x = 5-3a$, we have

$$2a+3b-12=0 \quad \text{and} \quad 5-3a=0.$$

Therefore $a = \dfrac{5}{3}$, $b = \dfrac{12-2a}{3} = \dfrac{26}{9}$.

Example 8. Find the integral value of k such that the equation $11x-2 = kx+15$ has positive integer solutions for x, and find such solutions.

Solution From the given equation we have

$$(11-k)x = 17.$$

Since it has at least one positive solution for x, so $k \neq 11$, and $x = \dfrac{17}{11-k}$. Since the fraction is an integer, $(11-k) \mid 17$, i.e. $k = -6$ or 10, and correspondingly, $x = 1$ or $x = 17$.

Example 9. Given that the equation $2a(x+6) = 4x+1$ has no solution, where a is a parameter, find the value of a.

Solution From the given equation $2a(x+6) = 4x+1$ we have $(2a-4)x = 1-12a$. Since it has no solution, this implies

$$2a-4=0 \qquad \text{and} \qquad 1-12a \neq 0,$$

therefore $a = 2$.

Example 10. Given that the equation $ax+4 = 3x-b$ has more than 1 solution for x. Find the value of $(4a+3b)^{2007}$.

Solution We rewrite the given equation in the form $(a-3)x = -(4+b)$. Then the equation has more than 1 solution implies that

$$a-3=0, \quad \text{and} \quad 4+b=0,$$

i.e. $a=3, b=-4$. Thus, $(4a+3b)^{2007} = 0^{2007} = 0$.

Testing Questions (A)

1. The equation taking -3 and 4 as its roots is
   (A) $(x-3)(x+4)=0$; (B) $(x-3)(x-4)=0$;
   (C) $(x+3)(x+4)=0$; (D) $(x+3)(x-4)=0$.

2. Given that the equation
   $$kx = 12$$
   has positive integer solution only, where k is an integer. Find the number of possible values of k.

3. The number of positive integers x satisfying the equation
   $$\frac{1}{x}+\frac{1}{x+1}+\frac{1}{x+2}=\frac{13}{12}$$
   is
   (A) 0 (B) 1 (C) 2 (D) infinitely many

4. Given that the solution of equation $3a-x=\frac{x}{2}+3$ is 4. Find the value of $(-a)^2-2a$.

5. Solve the equation $\frac{x-n}{m}-\frac{x-m}{n}=\frac{m}{n}$ (where $mn \neq 0$).

6. Solve the equation $[4ax-(a+b)](a+b)=0$, where a and b are constants.

7. Given that -2 is the solution of equation $\frac{1}{3}mx = 5x+(-2)^2$, find the value of the expression $(m^2-11m+17)^{2007}$.

8. Solve the equation $m^2x+1=m(x+1)$, where m is a parameter.

9. Given that k is a positive constant, and the equation $k^2x - k^2 = 2kx - 5k$ has a positive solution for x. Find the value of k.

10. Given that the equation $a(2x-1) = 3x - 3$ has no solution, find the value of the parameter a.

Testing Questions (B)

1. Given that the equations in x:

$$3\left[x - 2\left(x + \frac{a}{3}\right)\right] = 2x \quad \text{and} \quad \frac{3x+a}{3} - \frac{1+4x}{6} = 0$$

have a common solution. Find the common solution.

2. If positive numbers a, b, c satisfy $abc = 1$, solve the equation in x

$$\frac{2ax}{ab+a+1} + \frac{2bx}{bc+b+1} + \frac{2cx}{ca+c+1} = 1.$$

3. Given that the equation $\frac{8}{3}x - m = \frac{9}{4}x + 123$ has positive integer solution, where m is also a positive integer, find the minimum possible value of m.

4. Construct a linear equation with a constant term $-\frac{1}{2}$, such that its solution is equal to that of the equation $3[4x - (2x-6)] = 11x + 8$.

5. If $a_{n+1} = \dfrac{1}{1 + \dfrac{1}{a_n}}$ $(n = 1, 2, \ldots, 2008)$ and $a_1 = 1$, find the value of

$$a_1a_2 + a_2a_3 + a_3a_4 + \cdots + a_{2008}a_{2009}.$$

Lecture 4

System of Simultaneous Linear Equations

1. In general, the system of two equations of 2 variables can be expressed in the form
$$\begin{cases} a_1x + b_1y = c_1, \\ a_2x + b_2y = c_2. \end{cases}$$

2. To eliminating one variable for solving the system, we use (i) operations on equations as usual; (ii) substitution method. In many cases the method (i) is effective.

3. When $\frac{a_1}{a_2} \neq \frac{b_1}{b_2}$, the system has unique solution
$$x = \frac{c_1b_2 - c_2b_1}{a_1b_2 - a_2b_1}, \quad y = \frac{a_1c_2 - a_2c_1}{a_1b_2 - a_2b_1}.$$

4. When $\frac{a_1}{a_2} = \frac{b_1}{b_2} = \frac{c_1}{c_2}$ the system has two same equations, so it has infinitely many solutions, when $\frac{a_1}{a_2} = \frac{b_1}{b_2} \neq \frac{c_1}{c_2}$, the two equations are inconsistent, so it has no solution.

Examples

Example 1. Solve the system of equations
$$\begin{cases} \frac{x-y}{5} - \frac{x+y}{4} = \frac{1}{2}, \\ 2(x-y) - 3(x+y) + 1 = 0. \end{cases}$$

Solution **(I)** By operations on equations to eliminate a variable.

Simplifying the first equation, we have $4(x-y)-5(x+y)=10$, i.e.

$$x+9y=-10. \tag{4.1}$$

Simplifying the second equation, we have

$$x+5y=1. \tag{4.2}$$

By (4.1) – (4.2),

$$4y=-11, \quad \therefore y=-\frac{11}{4}.$$

From (4.2), $x=1-5y=1+\frac{55}{4}=\frac{59}{4}$. Thus, $x=\frac{59}{4}, y=-\frac{11}{4}$.

(II) By substitution to eliminate a variable.
From the first equation we have

$$x=-10-9y. \tag{4.3}$$

Substituting (4.3) into the second equation, we obtain

$$2(-10-9y-y)-3(-10-9y+y)+1=0,$$

$$4y=-11, \quad \therefore y=-\frac{11}{4}.$$

By substituting it back into (4.3), we obtain $x=-10+\frac{99}{4}=\frac{59}{4}$. Thus,

$$x=\frac{59}{4}, \quad y=-\frac{11}{4}.$$

Example 2. Solve the system of equations

$$5.4x+4.6y = 104, \tag{4.4}$$
$$4.6x+5.4y = 96. \tag{4.5}$$

Solution Notice the feature of coefficients, by (4.4) + (4.5), we obtain $10x+10y=200$, therefore

$$x+y=20. \tag{4.6}$$

By (4.4) – (4.5), it follows that $0.8x-0.8y=8$, therefore

$$x-y=10. \tag{4.7}$$

By $\frac{1}{2}((4.6)+(4.7))$ and $\frac{1}{2}((4.6)-(4.7))$ respectively, we obtain

$$x=15, \quad y=5.$$

Example 3. Solve the system of equations

$$\begin{cases} x+2(5x+y) &= 16, \\ 5x+y &= 7. \end{cases}$$

Solution By using 7 to substitute $5x+7$ in the first equation, we obtain $x+14=16$, therefore $x=2$.

Then from the second equation, $y=7-5x=-3$.

Note: The example indicates that not only a variable but an expression can be substituted also.

Example 4. Solve the system of equations

$$\begin{cases} \frac{x}{2}=\frac{y}{3}=\frac{z}{5}, \\ x+3y+6z=15. \end{cases}$$

Solution Let $t=\frac{x}{2}=\frac{y}{3}=\frac{z}{5}$, then

$$x=2t, y=3t, z=5t. \tag{4.8}$$

Substituting (4.8) into the second equation, we have $2t+9t+30t=15$, i.e. $t=\frac{15}{41}$. Thus

$$x=\frac{30}{41}, \quad y=\frac{45}{41}, \quad z=\frac{75}{41}.$$

Example 5. Solve the system of equations

$$\begin{cases} x+y &= 5 \\ y+z &= 6 \\ z+x &= 7. \end{cases}$$

Solution Let the given equations be labeled as

$$x+y = 5 \tag{4.9}$$

$$y+z = 6 \tag{4.10}$$

$$z+x = 7. \tag{4.11}$$

By $\frac{1}{2}$((4.9) + (4.10) + (4.11)), it follows that

$$x+y+z=9. \tag{4.12}$$

Then (4.12) – (4.9) yields $z = 4$;
(4.12) – (4.10) yields $x = 3$;
(4.12) – (4.11) yields $y = 2$.

Example 6. Solve the system of the equations

$$\begin{cases} x + 2y &= 5, \\ y + 2z &= 8, \\ z + 2u &= 11, \\ u + 2x &= 6. \end{cases}$$

Solution From the given equations we have the cyclic substitutions

$$x = 5 - 2y, \qquad y = 8 - 2z, \qquad z = 11 - 2u, \qquad u = 6 - 2x.$$

By substituting them sequentially, we have

$$\begin{aligned} x &= 5 - 2y = 5 - 2(8 - 2z) = -11 + 4z = -11 + 4(11 - 2u) = 33 - 8u \\ &= 33 - 8(6 - 2x) = -15 + 16x, \end{aligned}$$

therefore $x = 16x - 15$, i.e. $x = 1$, and then $u = 4$, $z = 3$, $y = 2$.

Example 7. Solve the system of equations

$$5x - y + 3z = a, \tag{4.13}$$
$$5y - z + 3x = b, \tag{4.14}$$
$$5z - x + 3y = c. \tag{4.15}$$

Solution By 2 × (4.13) + (4.14) – (4.15), it follows that

$$14x = 2a + b - c, \quad \therefore x = \frac{2a + b - c}{14}.$$

By 2 × (4.14) + (4.15) – (4.13), it follows that

$$14y = 2b + c - a, \quad \therefore y = \frac{2b + c - a}{14}.$$

Similarly, by 2 × (4.15) + (4.13) – (4.14), we have

$$14z = 2c + a - b, \quad \therefore z = \frac{2c + a - b}{14}.$$

Example 8. Given that x, y, z satisfy the system of equations

$$2000(x - y) + 2001(y - z) + 2002(z - x) = 0, \tag{4.16}$$
$$2000^2(x - y) + 2001^2(y - z) + 2002^2(z - x) = 2001, \tag{4.17}$$

find the value of $z - y$.

Solution Let $u = x - y, v = y - z, w = z - x$. Then u, v, w satisfy the following system of equations

$$u + v + w = 0, \quad (4.18)$$
$$2000u + 2001v + 2002w = 0, \quad (4.19)$$
$$2000^2u + 2001^2v + 2002^2w = 2001. \quad (4.20)$$

By 2001 × (4.18) – (4.19), we obtain

$$u - w = 0, \quad \text{i.e. } u = w.$$

From (4.18) again, we have $v = -2w$. By substituting it into (4.20), we have

$$(2000^2 - 2 \cdot 2001^2 + 2002^2)w = 2001,$$
$$[(2002 + 2001) - (2001 + 2000)]w = 2001,$$
$$2w = 2001, \qquad \therefore z - y = -v = 2w = 2001.$$

Example 9. Solve the system of equations for (x, y), and find the value of k.

$$x + (1 + k)y = 0, \quad (4.21)$$
$$(1 - k)x + ky = 1 + k, \quad (4.22)$$
$$(1 + k)x + (12 - k)y = -(1 + k). \quad (4.23)$$

Solution To eliminate k from the equation, by (4.22) + (4.23), we obtain

$$2x + 12y = 0, \quad \text{i.e. } x = -6y. \quad (4.24)$$

By substituting (4.24) into (4.21), we have $(k - 5)y = 0$. If $k \neq 5$, then $y = 0$ and so $x = 0$ also. From (4.22) we have $k = -1$.

If $k = 5$, (4.22) yields $(-4)(-6y) + 5y = 6$, so $y = \frac{6}{29}, x = -\frac{36}{29}$.

Testing Questions (A)

1. (CHINA/1997) Given that $x = 2, y = 1$ is the solution of system

$$\begin{cases} ax + by = 7, \\ bx + cy = 5, \end{cases}$$

then the relation between a and c is

(A) $4a + c = 9$; (B) $2a + c = 9$; (C) $4a - c = 9$; (D) $2a - c = 9$.

2. If the system in x and y

$$\begin{cases} 3x - y = 5, \\ 2x + y - z = 0, \\ 4ax + 5by - z = -22 \end{cases}$$

and the system in x and y

$$\begin{cases} ax - by + z = 8, \\ x + y + 5 = c, \\ 2x + 3y = -4 \end{cases}$$

have a same solution, then (a, b, c) is

(A) $(2, 3, 4)$; (B) $(3, 4, 5)$; (C) $(-2, -3, -4)$; (D) $(-3, -4, -5)$.

3. Determine the values of k such that the system of equations

$$\begin{cases} kx - y = -\frac{1}{3} \\ 3y = 1 - 6x \end{cases}$$

has unique solution, no solution, and infinitely many solutions respectively.

4. Given $\dfrac{ab}{a+b} = 2, \dfrac{ac}{a+c} = 5, \dfrac{bc}{b+c} = 4$, find the value of $a + b + c$.

5. Solve the system of equations

$$\begin{cases} x - y - z &= 5 \\ y - z - x &= 1 \\ z - x - y &= -15. \end{cases}$$

6. Solve the system of equations

$$\begin{cases} x - y + z &= 1 \\ y - z + u &= 2 \\ z - u + v &= 3 \\ u - v + x &= 4 \\ v - x + y &= 5. \end{cases}$$

7. Given

$$\frac{1}{x} + \frac{2}{y} + \frac{3}{z} = 0, \tag{4.25}$$

$$\frac{1}{x} - \frac{6}{y} - \frac{5}{z} = 0. \tag{4.26}$$

Find the value of $\dfrac{x}{y} + \dfrac{y}{z} + \dfrac{z}{x}$.

8. (CHINA/2001) Given that the system of equations

$$\begin{cases} mx + 2y = 10 \\ 3x - 2y = 0, \end{cases}$$

has an integer solution, i.e. x, y are both integers. Find the value of m^2.

9. As shown in the given figure, a, b, c, d, e, f are all rational numbers, such that the sums of three numbers on each row, each column and each diagonal are equal. Find the value of $a + b + c + d + e + f$.

a	b	6
c	d	e
f	7	2

10. Solve the system

$$x + y + z + u = 10, \tag{4.27}$$

$$2x + y + 4z + 3u = 29, \tag{4.28}$$

$$3x + 2y + z + 4u = 27, \tag{4.29}$$

$$4x + 3y + z + 2u = 22. \tag{4.30}$$

Testing Questions (B)

1. Given that the system of equations $\begin{cases} 3x + my = 7 \\ 2x + ny = 4 \end{cases}$ has no solution, where m, n are integers between -10 and 10 inclusive, find the values of m and n.

2. Solve the system of equations

$$\begin{cases} \dfrac{1}{x} + \dfrac{1}{y+z} = \dfrac{1}{2}, \\ \dfrac{1}{y} + \dfrac{1}{z+x} = \dfrac{1}{3}, \\ \dfrac{1}{z} + \dfrac{1}{x+y} = \dfrac{1}{4}. \end{cases}$$

3. Solve the system

$$\begin{cases} x(y+z-x) = 60 - 2x^2, \\ y(z+x-y) = 75 - 2y^2, \\ z(x+y-z) = 90 - 2z^2. \end{cases}$$

4. Find the values of a such that the system of equations in x and y

$$x + 2y = a + 6 \tag{4.31}$$

$$2x - y = 25 - 2a \tag{4.32}$$

has a positive integer solution (x, y).

5. Solve the system of equations

$$2x + y + z + u + v = 16, \tag{4.33}$$

$$x + 2y + z + u + v = 17, \tag{4.34}$$

$$x + y + 2z + u + v = 19, \tag{4.35}$$

$$x + y + z + 2u + v = 21, \tag{4.36}$$

$$x + y + z + u + 2v = 23. \tag{4.37}$$

Lecture 5

Multiplication Formulae

Basic Multiplication Formulae

(1) $$(a-b)(a+b)=a^2-b^2.$$

(2) $$(a \pm b)^2 = a^2 \pm 2ab + b^2.$$

(3) $$(a \pm b)(a^2 \mp ab + b^2) = a^3 \pm b^3.$$

Proof.

$$(a+b)(a^2-ab+b^2) = a^3 - a^2b + ab^2 + a^2b - ab^2 + b^3 = a^3 + b^3.$$

Use $(-b)$ to replace b in above formula, we obtain

$$(a-b)(a^2+ab+b^2) = a^3 - b^3.$$

□

(4) $$(a \pm b)^3 = a^3 \pm 3a^2b + 3ab^2 \pm b^3.$$

Proof.

$$\begin{aligned}(a+b)^3 &= (a+b)\cdot(a+b)^2 = (a+b)(a^2+2ab+b^2)\\ &= a^3 + 2a^2b + ab^2 + a^2b + 2ab^2 + b^3\\ &= a^3 + 3a^2b + 3ab^2 + b^3.\end{aligned}$$

Use $(-b)$ to replace b in above formula, we obtain

$$(a-b)^3 = a^3 - 3a^2b + 3ab^2 - b^3.$$

□

(5) $$(a+b+c)^2 = a^2 + b^2 + c^2 + 2ab + 2bc + 2ca.$$

Proof.

$$\begin{aligned}(a+b+c)^2 &= [(a+b)+c]^2 = (a+b)^2 + 2(a+b)c + c^2 \\ &= a^2 + 2ab + b^2 + 2ac + 2bc + c^2 \\ &= a^2 + b^2 + c^2 + 2ab + 2bc + 2ca.\end{aligned}$$

□

Generalization of Formulae

(1) $(a-b)(a^3 + a^2b + ab^2 + b^3) = a^4 - b^4.$

(2) $(a-b)(a^4 + a^3b + a^2b^2 + ab^3 + b^4) = a^5 - b^5.$

(3) $\boldsymbol{(a-b)(a^{n-1} + a^{n-2}b + \cdots + ab^{n-2} + b^{n-1}) = a^n - b^n}$ for all $n \in \mathbb{N}$.

Proof.

$$\begin{aligned}&(a-b)(a^{n-1} + a^{n-2}b + \cdots + ab^{n-2} + b^{n-1}) \\ &= (a^n + a^{n-1}b + \cdots + a^2b^{n-2} + ab^{n-1}) \\ &\quad -(a^{n-1}b + a^{n-2}b^2 + \cdots + ab^{n-1} + b^n) \\ &= a^n - b^n.\end{aligned}$$

□

(4) $\boldsymbol{(a+b)(a^{n-1} - a^{n-2}b + \cdots - ab^{n-2} + b^{n-1}) = a^n + b^n}$ for odd $n \in \mathbb{N}$.

Proof. For odd n, by using $(-b)$ to replace b in (3), we obtain

$$\begin{aligned}&(a+b)(a^{n-1} + a^{n-2}(-b) + a^{n-3}(-b)^2 + \cdots + a(-b)^{n-2} + (-b)^{n-1}) \\ &= a^n - (-b)^n,\end{aligned}$$

therefore

$$(a+b)(a^{n-1} - a^{n-2}b + a^{n-3}b^2 - \cdots - ab^{n-2} + b^{n-1}) = a^n + b^n.$$

□

(5) $$\begin{aligned}(a_1+a_2+\cdots+a_n)^2 = a_1^2+a_2^2+\cdots+a_n^2+2a_1a_2+2a_1a_3+\cdots+2a_1a_n \\ +2a_2a_3 + \cdots + 2a_2a_n + \cdots + 2a_{n-1}a_n.\end{aligned}$$

Proof.

$$\begin{aligned}&(a_1 + a_2 + \cdots + a_n)^2 \\ &= (a_1 + a_2 + a_3 + \cdots + a_n)(a_1 + a_2 + a_3 + \cdots + a_n) \\ &= a_1^2 + a_2^2 + \cdots + a_n^2 + 2a_1a_2 + 2a_1a_3 + \cdots + \cdots + 2a_1a_n \\ &\quad +2a_2a_3 + \cdots + 2a_2a_n + \cdots + 2a_{n-1}a_n.\end{aligned}$$

□

Derived Basic Formulae

(1) $a^2 + b^2 = (a \pm b)^2 \mp 2ab.$

(2) $(a + b)^2 - (a - b)^2 = 4ab.$

(3) $a^3 \pm b^3 = (a \pm b)^3 \mp 3ab(a \pm b).$

(4) $a^3 + b^3 + c^3 - 3abc = (a + b + c)(a^2 + b^2 + c^2 - ab - bc - ca).$

Proof.

$$\begin{aligned}
&a^3 + b^3 + c^3 - 3abc \\
&= (a^3 + 3a^2b + 3ab^2 + b^3) + c^3 - 3a^2b - 3ab^2 - 3abc \\
&= (a + b)^3 + c^3 - 3ab(a + b + c) \\
&= [(a + b) + c][(a + b)^2 - (a + b)c + c^2] - 3ab(a + b + c) \\
&= (a + b + c)(a^2 + 2ab + b^2 - ac - bc + c^2) - 3ab(a + b + c) \\
&= (a + b + c)(a^2 + b^2 + c^2 + 2ab - bc - ca - 3ab) \\
&= (a + b + c)(a^2 + b^2 + c^2 - ab - bc - ca).
\end{aligned}$$

□

Examples

Example 1. Evaluate the expression $(2 + 1)(2^2 + 1)(2^4 + 1) \cdots (2^{2^{10}} + 1) + 1.$

Solution By using the formula $(a - b)(a + b) = a^2 - b^2$ repeatedly, we have

$$\begin{aligned}
&(2 + 1)(2^2 + 1)(2^4 + 1) \cdots (2^{2^{10}} + 1) + 1 \\
&= (2 - 1)(2 + 1)(2^2 + 1)(2^4 + 1) \cdots (2^{2^{10}} + 1) + 1 \\
&= (2^2 - 1)(2^2 + 1)(2^4 + 1) \cdots (2^{2^{10}} + 1) + 1 \\
&= (2^4 - 1)(2^4 + 1) \cdots (2^{2^{10}} + 1) + 1 \\
&= \cdots = (2^{2^{10}} - 1)(2^{2^{10}} + 1) + 1 \\
&= ((2^{2^{10}})^2 - 1) + 1 = 2^{2 \cdot 2^{10}} = 2^{2^{11}} = 2^{2048}.
\end{aligned}$$

Example 2. Simplify the expression $(a^6 - b^6) \div (a^3 - b^3) \div (a^2 - ab + b^2).$

Solution By using the formulae $A^2-B^2=(A-B)(A+B)$ and $A^3+B^3=(A+B)(A^2-AB+B^2)$,

$$(a^6-b^6)\div(a^3-b^3)\div(a^2-ab+b^2)=\frac{a^6-b^6}{(a^3-b^3)(a^2-ab+b^2)}$$
$$=\frac{(a^3-b^3)(a^3+b^3)}{(a^3-b^3)(a^2-ab+b^2)}=\frac{a^3+b^3}{a^2-ab+b^2}$$
$$=\frac{(a+b)(a^2-ab+b^2)}{a^2-ab+b^2}=a+b.$$

Example 3. Given $x-y=8,\ xy=-15$, find the value of (i) $(x+y)^2$ and (ii) x^4+y^4.

Solution

(i) $$(x+y)^2=x^2+y^2+2xy=(x^2+y^2-2xy)+4xy=(x-y)^2+4xy$$
$$=8^2+4(-15)=4.$$

(ii) $$x^4+y^4=(x^4+2x^2y^2+y^4)-2x^2y^2=(x^2+y^2)^2-2(xy)^2$$
$$=[(x^2-2xy+y^2)+2xy]^2-2(-15)^2$$
$$=[(x-y)^2-30]^2-2(-15)^2=34^2-2(225)$$
$$=1156-450=706.$$

Example 4. Given $x+\frac{1}{x}=3$, find the value of (i) $x^3+\frac{1}{x^3}$; (ii) $x^4+\frac{1}{x^4}$.

Solution

(i) $$x^3+\frac{1}{x^3}=\left(x+\frac{1}{x}\right)\left(x^2+\frac{1}{x^2}-1\right)=3\left[\left(x+\frac{1}{x}\right)^2-3\right]$$
$$=3(3^2-3)=18.$$

(ii) $$x^4+\frac{1}{x^4}=\left[(x^2)^2+2+\left(\frac{1}{x^2}\right)^2\right]-2=\left(x^2+\frac{1}{x^2}\right)^2-2$$
$$=\left[\left(x+\frac{1}{x}\right)^2-2\right]^2-2=(3^2-2)^2-2=47.$$

Example 5. Given $x+y=\frac{5}{2}, x^2+y^2=\frac{13}{4}$, find the value of x^5+y^5.

Solution $(x^2+y^2)(x^3+y^3) = (x^5+y^5)+(xy)^2(x+y)$ implies that

$$(x^5+y^5) = \frac{13}{4}(x+y)(x^2+y^2-xy) - \frac{5}{2}(xy)^2 = \frac{65}{8}\left(\frac{13}{4}-xy\right) - \frac{5}{2}(xy)^2.$$

It suffices to find the value of xy. Then

$$xy = \frac{1}{2}[(x+y)^2-(x^2+y^2)] = \frac{1}{2}\left(\frac{25}{4}-\frac{13}{4}\right) = \frac{3}{2},$$

therefore

$$x^5+y^5 = \frac{65}{8}\left(\frac{13}{4}-\frac{3}{2}\right) - \frac{5}{2}\cdot\frac{9}{4} = \frac{455-180}{32} = \frac{275}{32}.$$

Example 6. Given that the real numbers x, y, z satisfy the system of equations

$$\begin{cases} x+y+z &= 6 \\ x^2+y^2+z^2 &= 26 \\ x^3+y^3+z^3 &= 90. \end{cases}$$

Find the values of xyz and $x^4+y^4+z^4$.

Solution $(x+y+z)^2 = (x^2+y^2+z^2)+2(xy+yz+zx)$ implies that

$$xy+yz+zx = \frac{1}{2}[(x+y+z)^2-(x^2+y^2+z^2)] = \frac{1}{2}[6^2-26] = 5.$$

Since $x^3+y^3+z^3-3xyz = (x+y+z)[(x^2+y^2+z^2-(xy+yz+zx)]$,

$$90-3xyz = 6[26-5] = 126,$$

$$\therefore xyz = \frac{1}{3}(90-126) = -12.$$

Further, by completing squares,

$$\begin{aligned} x^4+y^4+z^4 &= (x^2+y^2+z^2)^2 - 2(x^2y^2+y^2z^2+z^2x^2) \\ &= 26^2 - 2[(xy+yz+zx)^2 - 2(xy^2z+yz^2x+x^2yz)] \\ &= 26^2 - 2[5^2 - 2xyz(x+y+z)] \\ &= 26^2 - 2(25+24\cdot 6) = 676-338 = 338. \end{aligned}$$

Example 7. (SSSMO/2000) For any real numbers a, b and c, find the smallest possible values that the following expression can take:

$$3a^2+27b^2+5c^2-18ab-30c+237.$$

Solution By completing squares,

$$\begin{aligned}&3a^2+27b^2+5c^2-18ab-30c+237\\&=(3a^2-18ab+27b^2)+(5c^2-30c+45)+192\\&=3(a^2-6ab+9b^2)+5(c^2-6c+9)+192\\&=3(a-3b)^2+5(c-3)^2+192\ge 192.\end{aligned}$$

The value 192 is obtainable when $a=3b, c=3$. Thus, the smallest possible value of the given expression is 192.

Note: The technique for completing squares is an important tool for investigating extreme values of quadratic polynomials, here is an example.

Example 8. If $a,\ b,\ c,\ d>0$ and $a^4+b^4+c^4+d^4=4abcd$, prove that $a=b=c=d$.

Solution We rewrite the given equality in the form $a^4+b^4+c^4+d^4-4abcd=0$, and use the technique for completing squares, then

$$\begin{aligned}0&=a^4+b^4+c^4+d^4-4abcd\\&=(a^4-2a^2b^2+b^4)+(c^4-2c^2d^2+d^4)+(2a^2b^2+2c^2d^2-4abcd)\\&=(a^2-b^2)^2+(c^2-d^2)^2+2(ab-cd)^2,\end{aligned}$$

therefore $a^2-b^2=0, c^2-d^2=0, ab-cd=0$. Since $a,b,c,d>0$, so $a=b, c=d$, and $a^2=c^2$, i.e. $a=c$. Thus $a=b=c=d$.

Example 9. Given $a+b=c+d$ and $a^3+b^3=c^3+d^3$. Prove that $a^{2009}+b^{2009}=c^{2009}+d^{2009}$.

Solution $a+b=c+d$ yields $(a+b)^3=(c+d)^3$, therefore

$$a^3+3a^2b+3ab^2+b^3=c^3+3c^2d+3cd^2+d^3.$$
$$\because a^3+b^3=c^3+d^3,$$
$$\therefore 3a^2b+3ab^2=3c^2d+3cd^2,\quad \text{i.e. } 3ab(a+b)=3cd(c+d).$$

If $a+b=c+d=0$, then $b=-a, d=-c$, therefore

$$a^{2009}+b^{2009}=0=c^{2009}+d^{2009}.$$

If $a+b=c+d\neq 0$, then $ab=cd$, therefore

$$(a-b)^2=(a+b)^2-4ab=(c+d)^2-4cd=(c-d)^2.$$

(i) When $a-b=c-d$, considering $a+b=c+d$, it follows that $2a=2c$, i.e. $a=c$, and $b=d$ also.

(ii) When $a-b=-(c-d)$, considering $a+b=c+d$, it follows that $2a=2d$, i.e. $a=d$, and $b=c$ also.

The conclusion is true in each of the two cases.

Testing Questions (A)

1. (SSSMO/1998) Suppose a, b are two numbers such that
$$a^2 + b^2 + 8a - 14b + 65 = 0.$$
Find the value of $a^2 + ab + b^2$.

2. Given $a - b = 2, b - c = 4$, find the value of $a^2 + b^2 + c^2 - ab - bc - ca$.

3. For integers $a,\ b,\ c$ and d, rewrite the expression $(a^2 + b^2)(c^2 + d^2)$ as a sum of squares of two integers.

4. Given $14(a^2 + b^2 + c^2) = (a + 2b + 3c)^2$, find the ratio $a : b : c$.

5. Given $\dfrac{x}{x^2 + 3x + 1} = a\ (a \neq 0)$, find the value of $\dfrac{x^2}{x^4 + 3x^2 + 1}$.

6. Given $x + \dfrac{1}{x} = a$, find the value of $x^6 + \dfrac{1}{x^6}$ in terms of a.

7. Given that $a, b, c, d \neq 0$, and $a^4 + b^4 + c^4 + d^4 = 4abcd$. Prove that $a^2 = b^2 = c^2 = d^2$.

8 Given $a + b + c + d = 0$, prove that
$$a^3 + b^3 + c^3 + d^3 = 3(abc + bcd + cda + dab).$$

9. Given that $(x-2)^3+(y-2)^3+(z-2)^3 = 0, x^2+y^2+z^2 = 14, x+y+z = 6$, prove that at least one of x, y, z is 2.

10 Given that $a^3 + b^3 + c^3 = (a + b + c)^3$, prove that for any natural number n
$$a^{2n+1} + b^{2n+1} + c^{2n+1} = (a + b + c)^{2n+1}.$$

Testing Questions (B)

1. (CHNMOL/2005) If $M = 3x^2 - 8xy + 9y^2 - 4x + 6y + 13$ (where x, y are real numbers), then M must be

   (A) positive; (B) negative; (C) 0; (D) an integer.

2. Given $a + b = c + d$ and $a^2 + b^2 = c^2 + d^2$. Prove that $a^{2009} + b^{2009} = c^{2009} + d^{2009}$.

3. If $a+b+c=0$, prove that $2(a^4+b^4+c^4)=(a^2+b^2+c^2)^2$.

4. If $a+b=1, a^2+b^2=2$, find the value of a^7+b^7.

5. (CHNMOL/2004) Given that the real numbers a, b satisfy $a^3+b^3+3ab=1$, find $a+b$.

Lecture 6

Some Methods of Factorization

Basic Methods of Factorization

(I) **Extract the common factors from terms**: like

$$xm + ym + zm = m(x + y + z).$$

(II) **Apply multiplication formulae**: like those mentioned in Lecture 5. However, contrary to Lecture 5, at present each formula is applied for converting an expression in non-product form to a new expression in product form.

(III) **Cross-Multiplication**:

$$x^2 + (a + b)x + ab = (x + a)(x + b)$$

$$\begin{array}{ccc|c} x & \searrow\!\!\!\nearrow & a & ax \\ x & & b & bx \\ \hline & & & (a + b)x \end{array}$$

$$acx^2 + (ad + bc)x + bd = (ax + b)(cx + d).$$

$$\begin{array}{ccc|c} ax & \searrow\!\!\!\nearrow & b & bcx \\ cx & & d & adx \\ \hline & & & (ad + bc)x \end{array}$$

(IV) **By grouping, splitting, or inserting terms** to obtain common factors.

(V) **By substituting subexpressions** to simplify given expression.

(VI) **Coefficient-determining method**. First given the structure of the product, then determine the unknown parameters in the product by the comparison of coefficients.

(VII) **Factorization of symmetric or cyclic polynomials**. (cf. Lecture 15.)

Examples

Example 1. Factorize $(d^2 - c^2 + a^2 - b^2)^2 - 4(bc - da)^2$

Solution

$$(d^2 - c^2 + a^2 - b^2)^2 - 4(bc - da)^2 = (d^2 - c^2 + a^2 - b^2)^2 - (2bc - 2da)^2$$
$$= (d^2 - c^2 + a^2 - b^2 - 2bc + 2da)(d^2 - c^2 + a^2 - b^2 + 2bc - 2da)$$
$$= [(d + a)^2 - (b + c)^2] \cdot [(d - a)^2 - (b - c)^2]$$
$$= (d + a - b - c)(d + a + b + c)(d + b - a - c)(d + c - a - b).$$

Example 2. Factorize $64x^6 - 729y^{12}$.

Solution

$$64x^6 - 729y^{12} = (2x)^6 - (3y^2)^6 = [(2x)^3 - (3y^2)^3][(2x)^3 + (3y^2)^3]$$
$$= (2x - 3y^2)[(2x)^2 + (2x)(3y^2) + (3y^2)^2]$$
$$\cdot(2x + 3y^2)[(2x)^2 - (2x)(3y^2) + (3y^2)^2]$$
$$= (2x - 3y^2)(2x + 3y^2)(4x^2 + 6xy^2 + 9y^4)(4x^2 - 6xy^2 + 9y^4).$$

Example 3. Factor each of the following expressions:
(i) $2x^2 + x - 6$; (ii) $2x^2 - 10x + 8$.

Solution

$$x^2 + x - 6$$
$$= (x + 3)(x - 2).$$

x	3	$3x$
x	-2	$-2x$
		x

$$2x^2 - 10x + 8$$
$$= (2x - 2)(x - 4).$$

$2x$	-2	$-2x$
x	-4	$-8x$
		$-10x$

Example 4. Factorize $2a^3 + 6a^2 + 6a + 18$.

Solution

$$2a^3 + 6a^2 + 6a + 18 = 2[(a^3 + 3a^2 + 3a + 1) + 8] = 2[(a + 1)^3 + 2^3]$$
$$= 2(a + 3)[(a + 1)^2 - 2(a + 1) + 4] = 2(a + 3)(a^2 + 3).$$

Example 5. Factorize (i) $x^4 + 2x^3 + 7x^2 + 6x - 7$; (ii) $x^3 + 9x^2 + 23x + 15$.

Solution Let $y = x^2 + x$. Then

(i) $$x^4 + 2x^3 + 7x^2 + 6x - 7 = x^2(x^2 + x) + x(x^2 + x) + 6(x^2 + x) - 7$$
$$= (x^2 + x + 6)(x^2 + x) - 7 = y^2 + 6y - 7 = (y + 7)(y - 1)$$
$$= (x^2 + x + 7)(x^2 + x - 1).$$

(ii) $$x^3 + 9x^2 + 23x + 15 = x^2(x + 1) + 8x(x + 1) + 15(x + 1)$$
$$= (x + 1)(x^2 + 8x + 15) = (x + 1)(x + 3)(x + 5).$$

Example 6. Factorize (i) $(a+1)(a+2)(a+3)(a+4) - 120$; (ii) $x^5 + x + 1$.

Solution

(i) $$(a + 1)(a + 2)(a + 3)(a + 4) - 120$$
$$= [(a + 1)(a + 4)][(a + 2)(a + 3)] - 120$$
$$= (a^2 + 5a + 4)(a^2 + 5a + 6) - 120$$
$$= [(a^2 + 5a + 5) - 1][(a^2 + 5a + 5) + 1] - 120$$
$$= (a^2 + 5a + 5)^2 - 121 = (a^2 + 5a + 5)^2 - 11^2$$
$$= (a^2 + 5a - 6)(a^2 + 5a + 16) = (a - 1)(a + 6)(a^2 + 5a + 16).$$

(ii) $$x^5 + x + 1 = (x^5 - x^2) + (x^2 + x + 1)$$
$$= x^2(x^3 - 1) + (x^2 + x + 1)$$
$$= x^2(x - 1)(x^2 + x + 1) + (x^2 + x + 1)$$
$$= (x^2 + x + 1)[x^2(x - 1) + 1] = (x^2 + x + 1)(x^3 - x^2 + 1).$$

Example 7. Factorize $(2y - 3z)^3 + (3z - 4x)^3 + (4x - 2y)^3$.

Solution Let $2y - 3z = a$, $3z - 4x = b$, $4x - 2y = c$, then $a + b + c = 0$. Hence

$$(2y - 3z)^3 + (3z - 4x)^3 + (4x - 2y)^3 = a^3 + b^3 + c^3$$
$$= (a^3 + b^3 + c^3 - 3abc) + 3abc$$
$$= (a + b + c)(a^2 + b^2 + c^2 - bc - ca - ab) + 3abc$$
$$= 3abc = 3(2y - 3z)(3z - 4x)(4x - 2y).$$

Example 8. Factorize $(3a + 3b - 18ab)(3a + 3b - 2) + (1 - 9ab)^2$.

Solution The given expression is symmetric in $3a$ and $3b$, so we use $u =$

$3a+3b, v=(3a)(3b)$ to simplify the expression. Then

$$\begin{aligned}&(3a+3b-18ab)(3a+3b-2)+(1-9ab)^2=(u-2v)(u-2)+(1-v)^2\\&=u^2-2uv-2u+4v+v^2-2v+1=(u^2-2uv+v^2)+2(v-u)+1\\&=(v-u)^2+2(v-u)+1=(v-u+1)^2\\&=(9ab-3a-3b+1)^2=[(3a-1)(3b-1)]^2=(3a-1)^2(3b-1)^2.\end{aligned}$$

Note: Do not stop at $(v-u+1)^2$.

Example 9. Factorize $2x^2+7xy-4y^2-3x+6y-2$.

Solution Considering $2x^2+7xy-4y^2=(2x-y)(x+4y)$, let

$$2x^2+7xy-4y^2-3x+6y-2=(2x-y+a)(x+4y+b).$$

By expanding the product, it is obtained that

$$(2x-y+a)(x+4y+b)=2x^2+7xy-4y^2+(a+2b)x+(4a-b)y+ab.$$

By the comparison of coefficients, the following system of equations is obtained

$$a+2b = -3 \tag{6.1}$$
$$4a-b = 6, \tag{6.2}$$
$$ab = -2. \tag{6.3}$$

Then $2\times(6.2)+(6.1)$ yields $9a=9$, i.e. $a=1$. From (6.2), $b=6-4a=-2$. Since $(a,b)=(1,-2)$ satisfies (6.3), and it is the unique solution, we obtain

$$2x^2+7xy-4y^2-3x+6y-2=(2x-y+1)(x+4y-2).$$

Example 10. Given that $x^5-5qx+4r$ has a factor $(x-c)^2$ for some constant c. Prove that $q^5=r^4$.

Solution If $c=0$, then $x^2 \mid x^5-5qx+4r \Rightarrow r=q=0$, the conclusion is true. When $c\neq 0$, the condition means

$$\begin{aligned}&x^5-5qx+4r=(x^2-2cx+c^2)(x^3+ax^2+bx+d)\\&=x^5+(a-2c)x^4+(c^2+b-2ac)x^3+(ac^2-2bc+d)x^2\\&\quad+(bc^2-2cd)x+c^2d.\end{aligned}$$

Then the comparison of coefficients yields the following equations:

$$a=2c,\ \ b=2ac-c^2=3c^2,\ \ d=2bc-ac^2=4c^3=\frac{4r}{c^2}\Rightarrow r=c^5.$$

Further,

$$5q=2cd-bc^2=8c^4-3c^4=5c^4\Rightarrow q=c^4.$$

Thus, $q^5=c^{20}=r^4$.

Testing Questions (A)

1. Factorize the following expressions

(i) $x^9 + 7x^6y^3 + 7x^3y^6 + y^9$;
(ii) $4x^2 + y^2 + 9z^2 - 6yz + 12zx - 4xy$;
(iii) $(x^2 - 1)(x + 3)(x + 5) + 16$;
(iv) $(2x^2 - 4x + 1)^2 - 14x^2 + 28x + 3$;
(v) $x^3 - 3x^2 + (a + 2)x - 2a$;
(vi) $x^{11} + x^{10} + \cdots + x^2 + x + 1$.

2. Factorize the following expressions

(i) $x^4 - 2(a^2 + b^2)x^2 + (a^2 - b^2)^2$;
(ii) $(ab + 1)(a + 1)(b + 1) + ab$.

3. Prove that $81^6 - 9 \cdot 27^7 - 9^{11}$ is divisible by 45.

4. Prove that $\underbrace{44\cdots44}_{2n \text{ digits}} - \underbrace{88\cdots88}_{n \text{ digits}}$ is a perfect square number.

5. Factorize the following expressions
(i) $(x^2 + x - 1)^2 + x^2 + x - 3 = 0$;
(ii) $(x - y)^3 + (y - x - 2)^3 + 8$;
(iii) $(6x + 5)^2(3x + 2)(x + 1) - 6$;
(iv) $(x^2 + 5x + 6)(x^2 + 6x + 6) - 2x^2$;
(v) $(x^2 - 2x)^3 + (x^2 - 4x + 2)^3 - 8(x^2 - 3x + 1)^3$;
(vi) $a^3 + b^3 + c^3 + (a + b)(b + c)(c + a) - 2abc$.

6. Use the coefficient-determining method to factorize the following expressions.
(i) Given the expression $x^2 + xy - 2y^2 + 8x + ay - 9$. Find possible values of the constant a, such that the polynomial can be factorized as product of two linear polynomials.
(ii) $x^4 - x^3 + 4x^2 + 3x + 5$.

7. Given that $y^2 + 3y + 6$ is a factor of the polynomial $y^4 - 6y^3 + my^2 + ny + 36$. Find the values of constants m and n.

Testing Questions (B)

1. Factorize $2(x^2+6x+1)^2+5(x^2+1)(x^2+6x+1)+2(x^2+1)^2$.

2. (CHINA/2001) If x^2+2x+5 is a factor of x^4+ax^2+b, find the value of $a+b$.

3. Factor $(ab+cd)(a^2-b^2+c^2-d^2)+(ac+bd)(a^2+b^2-c^2-d^2)$.

4. Factorize $(ay+bx)^3+(ax+by)^3-(a^3+b^3)(x^3+y^3)$.

5. Given that a, b, c are three distinct positive integers. Prove that among the numbers
$$a^5b-ab^5,\ b^5c-bc^5,\ c^5a-ca^5,$$
there must be one that is divisible by 8.

Lecture 7

Absolute Value and Its Applications

For any real number a, we define its **absolute value**, denoted by $|a|$, as follows:

$$|a| = \begin{cases} a, & \text{if } a > 0 \\ 0, & \text{if } a = 0 \\ -a, & \text{if } a < 0. \end{cases}$$

Geometrically, any real number a is denoted by a point on the number axis, and the absolute value of a is the distance of the point representing a from the origin of the number axis.

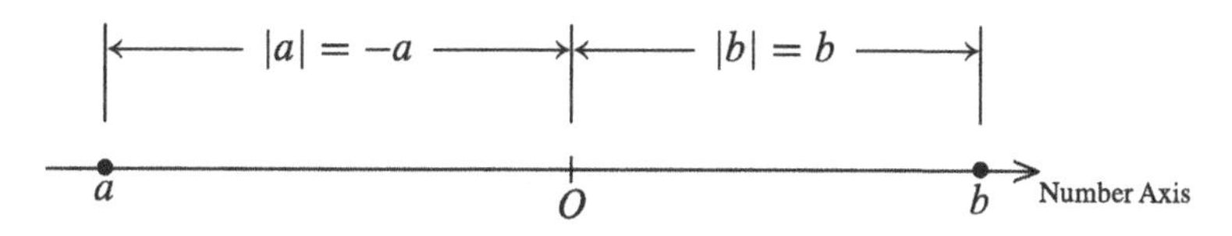

More general, the expression $|a - b|$ denotes the distance between the points on the number axis representing the numbers a and b.

When taking absolute value to any algebraic expression, a non-negative value can be obtained always from it by eliminating its negative sign if the value of the expression is "–". This rule is similar to that of taking square to that expression.

Basic Properties of Absolute Value

1. $|a| = |-a|$;
2. $-|a| \le a \le |a|$;
3. $|a| = |b|$ if and only if $a = b$ or $a = -b$.
4. $|a^n| = |a|^n$ for any positive integer n;
5. $|ab| = |a| \cdot |b|$;

6. $\left|\frac{a}{b}\right| = \frac{|a|}{|b|}$ if $b \neq 0$;
7. $|a \pm b| \leq |a| + |b|$.

Examples

Example 1. Is there a real number x such that $\frac{|x-|x||}{x}$ is a positive number?

Solution It is clear that $x \neq 0$.
For $x > 0$, $\frac{|x-|x||}{x} = \frac{|x-x|}{x} = \frac{0}{x} = 0$.
For $x < 0$, $\frac{|x-|x||}{x} = \frac{|x-(-x)|}{x} = \frac{|2x|}{x} = \frac{-2x}{x} = -2$.
Thus, there is no real number x such that the given fraction is positive.

Example 2. If a, b, c are non-zero real numbers, find all possible values of the expression $\frac{a}{|a|} + \frac{b}{|b|} + \frac{c}{|c|}$.

Solution Since $\frac{x}{|x|} = 1$ for any $x > 0$ and $\frac{x}{|x|} = -1$ for any $x < 0$,

$$\frac{a}{|a|} + \frac{b}{|b|} + \frac{c}{|c|} = -3 \text{ if } a, b, c \text{ are all negative;}$$
$$\frac{a}{|a|} + \frac{b}{|b|} + \frac{c}{|c|} = -1 \text{ if exactly two of } a, b, c \text{ are negative;}$$
$$\frac{a}{|a|} + \frac{b}{|b|} + \frac{c}{|c|} = 1 \text{ if exactly one of } a, b, c \text{ is negative;}$$
$$\frac{a}{|a|} + \frac{b}{|b|} + \frac{c}{|c|} = 3 \text{ if } a, b, c \text{ are all positive.}$$

Thus, the possible values of the given expression are $-3, -1, 1$ and 3.

Example 3. Determine the condition for the equality $\left|\frac{a-b}{a}\right| = \frac{b-a}{a}$.

Solution The given equality implies that $a \neq 0$ and $\left|\frac{a-b}{a}\right| = -\frac{a-b}{a}$, therefore $\frac{a-b}{a} \leq 0$. Since

$$\frac{a-b}{a} \leq 0 \Longleftrightarrow 1 - \frac{b}{a} \leq 0 \Longleftrightarrow \frac{b}{a} \geq 1,$$

the condition on a and b is $\frac{b}{a} \geq 1$.

Example 4. a, b, c are real numbers satisfying $(3a+6)^2 + |\frac{1}{4}b - 10| + |c+3| = 0$. Find the value of $a^{10} + bc$.

Solution Each term on the left hand side of the given equality is non-negative, we must have

$$3a + 6 = 0 \quad \frac{b}{4} - 10 = 0 \qquad c + 3 = 0$$

at the same time, therefore $a = -2, b = 40, c = -3$, so that

$$a^{10} + bc = (-2)^{10} + (40)(-3) = 1024 - 120 = 904.$$

Example 5. Given $1 < x < 3$, simplify the following expressions:

(i) $\frac{|x-3|}{x-3} - \frac{|x-1|}{(1-x)}$. (ii) $|x-1| + |3-x|$.

Solution For simplifying an expression with absolute values, it is needed to convert it to a normal expression by removing the absolute signs. For this, we need to *partition the range of x into several intervals*, so that on each interval the sign of the expression is fixed (only positive or only negative). For example, for removing the absolute signs of $|x-3|$, it is needed to take $x - 3 = 0$, i.e. $x = 3$ as the origin, and the sign of $x - 3$ changes at this point: it is positive when $x > 3$, and negative when $x < 3$, so it is needed to discuss $|x-3|$ for $x > 3$ and $x < 3$ separately. Thus, since the range of x is right to 1 and left to 3,

(i) $x - 3 < 0$ and $x - 1 > 0$ implies $|x-3| = -(x-3), |x-1| = x - 1$, therefore

$$\frac{|x-3|}{x-3} - \frac{|x-1|}{(1-x)} = \frac{-(x-3)}{x-3} - \frac{x-1}{1-x} = -1 - (-1) = 0.$$

(ii) By the same reason, $|x-1| + |3-x| = (x-1) + (3-x) = 2$.

Example 6. Given $1 < x < 3$, simplify $|x-2| + 2|x|$.

Solution The zero points of $|x-2|$ and $|x|$ are $x = 2$ and $x = 0$ respectively, it is needed to partition the range of x into two intervals: $1 < x \leq 2$ and $2 < x < 3$.

(i) When $1 < x \leq 2$, $|x-2| + 2|x| = 2 - x + 2x = 2 + x$;

(ii) when $2 < x < 3$, $|x-2| + 2|x| = x - 2 + 2x = 3x - 2$.

Example 7. Simplify $||x+2| - 7| - |7 - |x-5||$ for $-2 < x < 5$.

Solution We remove the absolute signs from inner layers to outer layers. Since $x + 2 > 0$ and $x - 5 < 0$,

$$||x+2|-7|-|7-|x-5|| = |(x+2)-7|-|7-(5-x)|$$
$$= |x-5|-|2+x| = (5-x)-(2+x) = 3-2x.$$

Example 8. (AHSME/1990) Determine the number of real solutions of the equation $|x-2|+|x-3|=1$.

(A) 0 (B) 1 (C) 2 (D) 3 (E) more than 3.

Solution We need to discuss Three cases: $x \le 2$; $2 < x \le 3$, and $3 < x$.

(i) When $x \le 2$,

$$|x-2|+|x-3|=1 \Leftrightarrow (2-x)+(3-x)=1 \Leftrightarrow x=2;$$

(ii) When $2 < x \le 3$,

$|x-2|+|x-3|=1 \Leftrightarrow (x-2)+(3-x)=1 \Leftrightarrow$ any $x \in (2,3]$ is a solution.

Thus, the answer is (E).

Example 9. Let the positions of points on the number axis representing real numbers a, b, c be as shown in the following diagram. Find the value of the expression

$$|b-a|-|a-c|+|c-b|.$$

Solution From the diagram we find that $c < b < 0 < a < -c$, therefore

$$|b-a|-|a-c|+|c-b| = (a-b)-(a-c)+b-c = 0.$$

Thus, the value of the expression is 0.

Example 10. Given $m = |x+2|+|x-1|-|2x-4|$. Find the maximum value of m.

Solution We discuss the maximum value of m on each of the following four intervals.

(i) When $x \le -2$, then

$$m = -(x+2)-(x-1)+(2x-4) = -5.$$

(ii) When $-2 < x \le 1$, then

$$m = (x+2)-(x-1)+(2x-4) = 2x-1 \le 1.$$

(iii) When $1 < x \le 2$, then

$$m = (x+2) + (x-1) + (2x-4) = 4x - 3 \le 5.$$

(iv) When $2 < x$, then

$$m = (x+2) + (x-1) + (4-2x) = 5.$$

Thus, $m_{\max} = 5$.

Example 11. Let $a < b < c$, Find the minimum value of the expression

$$y = |x-a| + |x-b| + |x-c|.$$

Solution

(i) When $x \le a$,

$$y = (a-x) + (b-x) + (c-x) \ge (b-a) + (c-a).$$

(ii) When $a < x \le b$,

$$y = (x-a)+(b-x)+(c-x) = (b-a)+(c-x) \ge (b-a)+(c-b) = c-a.$$

(iii) When $b < x \le c$,

$$y = (x-a)+(x-b)+(c-x) = (x-a)+(c-b) > (b-a)+(c-b) = c-a.$$

(iv) When $c < x$,

$$y = (x-a) + (x-b) + (x-c) > (b-a) + (c-b) + (x-c) > c-a.$$

Thus, $y_{\min} = c - a$, and y reaches this minimum value at $x = b$.

Testing Questions (A)

1. Simplify $\dfrac{|x + |x||}{x}$.

2. Given that $\dfrac{2x-1}{3} - 1 \ge x - \dfrac{5-3x}{2}$. Find the maximum and minimum values of the expression $|x-1| - |x+3|$.

3. If real number x satisfies the equation $|1-x| = 1 + |x|$, then $|x-1|$ is equal to

   (A) 1, (B) $-(x-1)$, (C) $x-1$, (D) $1-x$.

4. What is the minimum value of $|x+1|+|x-2|+|x-3|$?

5. If $x<0$, find the value of $\frac{||x|-2x|}{3}$.

6. Given $a<b<c<d$, find the minimum value of $|x-a|+|x-b|+|x-c|+|x-d|$.

7. If two real numbers a and b satisfy $|a+b|=a-b$, find the value of ab.

8. Given that a,b,c are integers. If $|a-b|^{19}+|c-a|^{19}=1$, find the value of $|c-a|+|a-b|+|b-c|$.

9. Given $a=2009$. Find the value of $|2a^3-3a^2-2a+1|-|2a^3-3a^2-3a-2009|$.

10. a,b are two constants with $|a|>0$. If the equation $||x-a|-b|=3$ has three distinct solutions for x, find the value of b.

Testing Questions (B)

1. Given that n real numbers $x_1,x_2,\ldots,x_n$ satisfy $|x_i|<1$ $(i=1,2,\ldots,n)$, and

$$|x_1|+|x_2|+\cdots+|x_n|=49+|x_1+x_2+\cdots+x_n|.$$

Find the minimum value of n.

2. Given that $a_1<a_2<\cdots<a_n$are constants, find the minimum value of $|x-a_1|+|x-a_2|+\cdots+|x-a_n|$.

3. When $2x+|4-5x|+|1-3x|+4$takes some constant value on some interval, find the interval and the constant value.

4. Given that real numbers a,b,c are all not zero, and $a+b+c=0$. Find the value of $x^{2007}-2007x+2007$, where $x=-\left|\frac{|a|}{b+c}+\frac{|b|}{a+c}+\frac{|c|}{a+b}\right|$.

5. The numbers $1,2,3,\cdots,199,200$ are partitioned into two groups of 100 each, and the numbers in one group are arranged in ascending order: $a_1<a_2<a_3<\cdots<a_{100}$, and those in the other group are arranged in descending order: $b_1>b_2>b_3>\cdots>b_2>b_1$. Find the value of the expression

$$|a_1-b_1|+|a_2-b_2|+\cdots+|a_{99}-b_{99}|+|a_{100}-b_{100}|.$$

Lecture 8

Linear Equations with Absolute Values

To solve a linear equation with absolute values, we need to remove the absolute value signs in the equation.

In the simplest case $|P(x)| = Q(x)$, where $P(x)$, $Q(x)$ are two expressions with $Q(x) \geq 0$, by the properties of absolute values, we can remove the absolute value signs by using its equivalent form

$$P(x) = Q(x) \quad \text{or} \quad P(x) = -Q(x) \quad \text{or} \quad (P(x))^2 = (Q(x))^2.$$

If there are more then one pair of absolute signs in a same layer, like $|ax + b| - |cx + d| = e$, it is needed to partition the range of the variable x into several intervals to discuss (cf. Lecture 7).

Examples

Example 1. Solve equation $|3x + 2| = 4$.

Solution To remove the absolute signs from $|3x + 2| = 4$ we have

$$|3x + 2| = 4 \iff 3x + 2 = -4 \text{ or } 3x + 2 = 4,$$
$$\iff 3x = -6 \quad \text{or} \quad 3x = 2,$$
$$x = -2 \qquad \text{or} \qquad x = \frac{2}{3}.$$

Example 2. Solve equation $|x - |3x + 1|| = 4$.

Solution For removing multiple layers of absolute value signs, we remove them layer by layer from outer layer to inner layer. From the given equation we have $x - |3x + 1| = 4$ or $x - |3x + 1| = -4$.

From the first equation $x - |3x + 1| = 4$ we have

$$x-|3x+1| = 4 \iff 0 \leq x-4 = |3x+1| \iff -x+4 = 3x+1 \text{ or } x-4 = 3x+1,$$

therefore $x = \frac{3}{4}$ or $x = -\frac{5}{2}$, which contradict the requirement $x \geq 4$, so the two solutions are not acceptable.

The second equation $x - |3x+1| = -4$ implies that

$$x - |3x+1| = -4 \Longleftrightarrow 0 \leq x+4 = |3x+1|$$
$$\Longleftrightarrow -x-4 = 3x+1 \text{ or } x+4 = 3x+1,$$

therefore $x_1 = -\frac{5}{4}$, $x_2 = \frac{3}{2}$.

Example 3. Solve equation $|||x|-2|-1| = 3$.

Solution There are three layers of absolute values. Similar to Example 2,

$$|||x|-2|-1| = 3 \Longleftrightarrow ||x|-2|-1 = 3 \text{ or } ||x|-2|-1 = -3$$
$$\Longleftrightarrow ||x|-2| = 4 \text{ or } ||x|-2|-1| = -2 \text{ (no solution)}$$
$$\Longleftrightarrow |x|-2 = 4 \text{ or } |x|-2 = -4 \Longleftrightarrow |x| = 6 \text{ or } |x| = -2 \text{ (no solution)}$$
$$\Longleftrightarrow |x| = 6 \Longleftrightarrow x_1 = 6, \quad x_2 = -6.$$

Example 4. If $|x-2| + x - 2 = 0$, then the range of x is
(A) $x > 2$, (B) $x < 2$, (C) $x \geq 2$, (D) $x \leq 2$.

Solution The given equation produces $|x-2| = 2-x$, so $x \leq 2$ and

$$|x-2| = 2-x \Longleftrightarrow x-2 = 2-x \text{ or } x-2 = x-2 \Longleftrightarrow x = 2 \text{ or } x \leq 2.$$

Since $x = 2$ is contained by the set $x \leq 2$, the answer is $x \leq 2$, i.e. (D).

Example 5. If $||4m+5| - b| = 6$ is an equation in m, and it has three distinct solutions, find the value of the rational number b.

Solution From the given equation we have (i) $|4m+5| - b = 6$ or (ii) $|4m+5| - b = -6$.

If (i) has exactly one solution, then $b + 6 = 0$, i.e. $b = -6$ which implies (ii) should be $|4m+5| = -12$, so no solutions. Thus $b \neq -6$ and (i) has two solutions but (ii) has exactly one solution, so $b - 6 = 0$, i.e. $b = 6$. In fact, when $b = 6$ then (i) becomes $|4m+5| = 12$,

$$4m+5 = 12 \quad \text{or} \quad 4m+5 = -12,$$
$$m = \frac{7}{4} \text{ or } m = -\frac{17}{4},$$

and, from (ii) the third root $m = -\frac{5}{4}$.

Example 6. Solve equation $|x-1| + 2|x| - 3|x+1| - |x+2| = x$.

Solution Letting each of $|x-1|, |x|, |x+1|, |x+2|$ be 0, we get $x = 1, 0, -1, -2$. By using these four points as partition points, the number axis is partitioned as five intervals: $x \le -2,\ -2 < x \le -1,\ -1 < x \le 0,\ 0 < x \le 1,\ 1 < x$.

(i) When $x \le -2$, then
$(1-x) + 2(-x) + 3(x+1) + (x+2) = x \iff 6 = 0, \ \therefore$ no solution;

(ii) when $-2 < x \le -1$, then
$(1-x) + 2(-x) + 3(x+1) - (x+2) = x, \iff x = 1 \ \therefore$ no solution;

(iii) when $-1 < x \le 0$, then
$1 - x + 2(-x) - 3(x+1) - (x+2) = x, \iff 8x = -4, \ \therefore x = -\frac{1}{2};$

(iv) when $0 < x \le 1$, then
$(1-x) + 2x - 3(x+1) - (x+2) = x \iff 4x = -4, \ \therefore x = -1,$
therefore no solution;

(v) when $1 < x$, then
$(x-1) + 2x - 3(x+1) - (x+2) = x \iff 2x = -6 \ x = -3$
therefore no solution.

Thus $x = -\frac{1}{2}$ is the unique solution.

Example 7. If $|x+1| + (y+2)^2 = 0$ and $ax - 3ay = 1$, find the value of a.

Solution Since $|x+1| \ge 0$ and $(y+2)^2 \ge 0$ for any real x, y, so $x+1 = 0$ and $y+2 = 0$, i.e. $x = -1, y = -2$. By substituting them into $ax - 3ay = 1$, it follows that $-a + 6a = 1$, therefore $a = \frac{1}{5}$.

Example 8. How many pairs (x, y) of two integers satisfy the equation $|xy| + |x-y| = 1$?

Solution $|xy| \ge 0$ and $|x-y| \ge 0$ implies that

(i)| $|xy| = 1, |x-y| = 0$ or (ii) $|xy| = 0, |x-y| = 1$.

(i) implies that $x = y$ and $x^2 = y^2 = 1$, so its solutions (x, y) are $(1, 1)$ or $(-1, -1)$.

(ii) implies that at least one of x, y is 0. When $x = 0$, then $|y| = 1$, i.e. $y = \pm 1$; if $y = 0$, then $|x| = 1$, i.e. $x = \pm 1$. Hence the four solutions for (x, y) are $(0, 1),\ (0, -1),\ ((1, 0),\ (-1, 0)$.

Thus there are 6 solutions for (x, y) in total.

Example 9. If $|x+1| - |x-3| = a$ is an equation in x, and it has infinitely many solutions, find the value of a.

Solution By $-1, 3$ partition the number axis into three parts: $x \le -1$, $-1 < x \le 3$, $3 < x$.

(i) When $x \le -1$, then
$-(x+1) - (3-x) = a \iff -4 = a$. Therefore any value of x not greater than -1 is a solution if $a = -4$.

(ii) When $-1 < x \le 3$, then
$(x+1) - (3-x) = a \iff 2x = a + 2 \iff x = \frac{1}{2}(a+2)$, i.e. the solution is unique if any.

(iii) When $3 < x$, then
$(x+1) - (x-3) = a \iff 4 = a$. Therefore any value of x greater than 3 is a solution if $a = 4$.

Thus, the possible values of a are -4 and 4.

Example 10. (AHSME/1988) If $|x| + x + y = 10$ and $x + |y| - y = 12$, find $x + y$.

Solution There are four possible cases: (i) $x, y > 0$; (ii) $x, y \le 0$; (iii) $x > 0, y \le 0$ and (iv) $x \le 0, y > 0$.

(i) If $x > 0, y > 0$ then $2x + y = 10, x = 12 \iff y < 0$, a contradiction, so no solution;

(ii) If $x \le 0$ and $y \le 0$, then $y = 10$, a contradiction, so no solution;

(iii) If $x > 0, y \le 0$, then $2x + y = 10, x - 2y = 12$. By eliminating y, we have $x = \dfrac{32}{5}$, so $y = -\dfrac{14}{5}$.

Thus, $x + y = \dfrac{18}{5}$.

Testing Questions (A)

1. Solve equation $|5x - 4| - 2x = 3$.

2. (CHINA/2000) a is an integer satisfying the equation $|2a+7|+|2a-1| = 8$. Then the number of solutions for a is

   (A) 5 (B) 4 (C) 3 (D) 2.

3. (AHSME/1984) The number of distinct solutions of the equation $|x - |2x + 1|| = 3$ is

   (A) 0, (B) 1, (C) 2, (D) 3, (E) 4.

4. (CHNMOL/1987) Given that the equation $|x| = ax + 1$ has exactly one negative solution and has no positive solution. then the range of a is

   (A) $a > 1$, (B) $a = 1$, (C) $a \geq 1$, (D) none of preceding.

5. (CHNMOL/1986) If the equation $||x - 2| - 1| = a$ has exactly three integer solution for x, then the value of a is

   (A) 0, (B) 1, (C) 2, (D) 3.

6. If the equation $\frac{a}{2008}|x| - x - 2008 = 0$ has only negative solutions for x, find the range of a.

7. In the equations in x

   (i) $|3x - 4| + 2m = 0$, (ii) $|4x - 5| + 3n = 0$, (iii) $|5x - 6| + 4k = 0$,

   m, n, k are constants such that (i) has no solution, (ii) has exactly one solution, and (iii) has two solutions. Then

   (A) $m > n > k$, (B) $n > k > m$, (C) $k > m > n$, (D) $m > k > n$.

8. Solve the system of simultaneous equations

$$\begin{cases} |x - y| = x + y - 2, \\ |x + y| = x + 2. \end{cases}$$

9. (AHSME/1958) We may say concerning the solution of $|x|^2 + |x| - 6 = 0$ that:

   (A) there is only one root; (B) the sum of the roots is 1;

   (C) the sum of the roots is 0; (D) the product of the roots is $+4$;

   (E) the product of the roots is -6.

10. (CHINA/2001) Solve the system $x + 3y + |3x - y| = 19, 2x + y = 6$.

Testing Questions (B)

1. Solve the system

$$\begin{cases} |x - 2y| = 1, \\ |x| + |y| = 2. \end{cases}$$

2. (CHINA/1990) For the equation with 4 layers of absolute value signs $||||x-1|-1|-1|-1|=0$,

   (A) the solutions are $0, 2, 4$; (B) $0, 2, 4$ are not solutions;

   (C) the solutions are within the three values $0, 2, 4$;

   (D) $0, 2, 4$ are not all of the solutions.

3. (CHNMOL/1995) Given that a, b are real numbers satisfying the inequality $||a|-(a+b)| < |a-|a+b||$, then

   (A) $a > 0, b > 0$ (B) $a < 0, b > 0$ (C) $a > 0, b < 0$ (D) $a < 0, b < 0$.

4. Given that $\dfrac{1}{|x-2|} = \dfrac{1}{|x-52a|}$ is an equation in x,

   (i) solve the equation, (ii) prove that the solutions must be composite numbers if a is the square of an odd prime number.

5. (IMO/1966) Solve the system of equations

$$\begin{array}{llllr} & |a_1-a_2|\,x_2 & +|a_1-a_3|\,x_3 & +|a_1-a_4|\,x_4 & =1 \\ |a_2-a_1|\,x_1 & & +|a_2-a_3|\,x_3 & +|a_2-a_4|\,x_4 & =1 \\ |a_3-a_1|\,x_1 & +|a_3-a_2|\,x_2 & & +|a_3-a_4|\,x_4 & =1 \\ |a_4-a_1|\,x_1 & +|a_4-a_2|\,x_2 & +|a_4-a_3|\,x_3 & & =1 \end{array}$$

   where a_1, a_2, a_3, a_4 are four different real numbers.

Lecture 9

Sides and Angles of a Triangle

Basic Knowledge

1. For any triangle, the sum of lengths of any two sides must be longer than the length of the third side. In other words, the difference of lengths of any two sides must be less than the length of the third side.
2. The sum of three interior angles of a triangle is $180°$.
3. The sum of all interior angles of an n-sided polygon is $(n-2) \times 180°$.
4. The sum of all exterior angles of a convex n-sided polygon is $360°$.
5. An exterior angle of a triangle is equal to the sum of the two opposite interior angles.
6. For a triangle, the opposite side of a bigger interior angle is longer than that of a smaller angle, and *vice versa*.
7. For triangles ABC and $A_1B_1C_1$, if $AB = A_1B_1, CB = C_1B_1$, then $AC > A_1C_1$ if and only if $\angle ABC > \angle A_1B_1C_1$.

Examples

Example 1.

(1) When each side of a triangle has a length which is a prime factor of 2001, how many different such triangles are there?

(2) How many isosceles triangles are there, such that each of its sides has an integral length, and its perimeter is 144?

Solutions (1) Since $2001 = 3 \times 23 \times 29$, The triangles with sides of the following lengths exist:

$$\{3,3,3\}; \quad \{23,23,23\}; \quad \{29,29,29\};$$

$$\{3,23,23\}; \quad \{3,29,29\}; \quad \{23,29,29\}; \quad \{23,23,29\}.$$

There are 7 possible triangles in total.

(2) Suppose that each leg of the isosceles triangle has length n, then its base has a length $144 - 2n = 2(72 - n)$, i.e. the length of the base must be even.

(i) If $n \geq 144 - 2n$ i.e. $3n \geq 144$, then $n \geq 48$. Since $2n \leq 144 - 2 = 142$, i.e. $n \leq 71$, we have $48 \leq n \leq 71$, there are 24 possible values for n.

(ii) If $n < 144 - 2n$, then $n < 48$. From triangle inequality, $2n > 144 - 2n$, i.e. $n > 36$, then $36 < n < 48$, so n has $47 - 36 = 11$ possible values.

Thus, there are together $24 + 11 = 35$ possible isosceles triangles.

Example 2. Given a convex polygon of which the sum of all interior angles excluding one is 2200°. Find the excluded interior angle.

Solution Since the sum of interior angle of a n-sided convex polygon is $(n-2) \cdot 180°$, from

$$2200° = 12 \times 180° + 40° = 13 \times 180° - 140°,$$

it follows that that $n = 13 + 2 = 15$, and the excluded angle is 140°.

Example 3. As shown in the diagram below, in $\triangle ABC$, $\angle B > \angle C$, AD is the bisector of the $\angle BAC$, $AE \perp BC$ at E. Prove that $\angle DAE = \frac{1}{2}(\angle B - \angle C)$.

Solution Since

$$\begin{aligned}
\angle BAD &= \tfrac{1}{2}\angle BAC = \tfrac{1}{2}(180° - \angle B - \angle C) \\
&= 90° - \tfrac{1}{2}\angle B - \tfrac{1}{2}\angle C, \\
\angle DAE &= \angle BAD - \angle BAE \\
&= \angle BAD - (90° - \angle B) \\
&= 90° - \tfrac{1}{2}(\angle B + \angle C) - 90° + \angle B \\
&= \tfrac{1}{2}(\angle B - \angle C).
\end{aligned}$$

A

B E D C

Example 4. (AHSME/1956) In the figure below $AB = AC$, $\angle BAD = 30°$, and $AE = AD$. Then $\angle CDE$ equals:

(A) 7.5°, (B) 10°, (C) 12.5°, (D) 15°, (E) 20°.

Solution Let $\angle CDE = x$, then

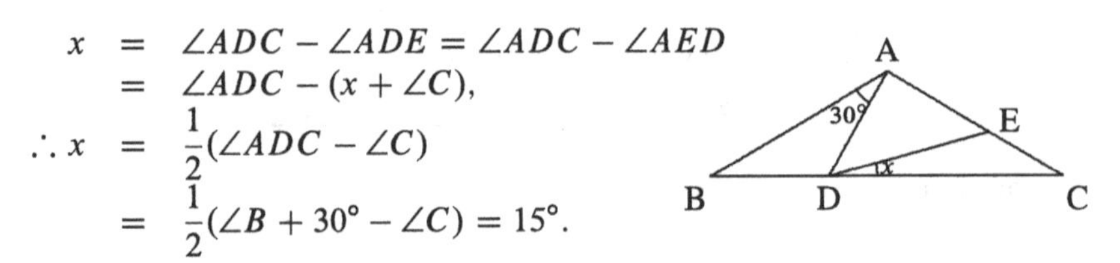

$$\begin{aligned}
x &= \angle ADC - \angle ADE = \angle ADC - \angle AED \\
&= \angle ADC - (x + \angle C), \\
\therefore x &= \frac{1}{2}(\angle ADC - \angle C) \\
&= \frac{1}{2}(\angle B + 30° - \angle C) = 15°.
\end{aligned}$$

Example 5. As shown in the figure, $AB = BC = CD = DE = EF = FG = GH$, $\angle\alpha = 70°$. Find the size of $\angle\beta$ in degrees.

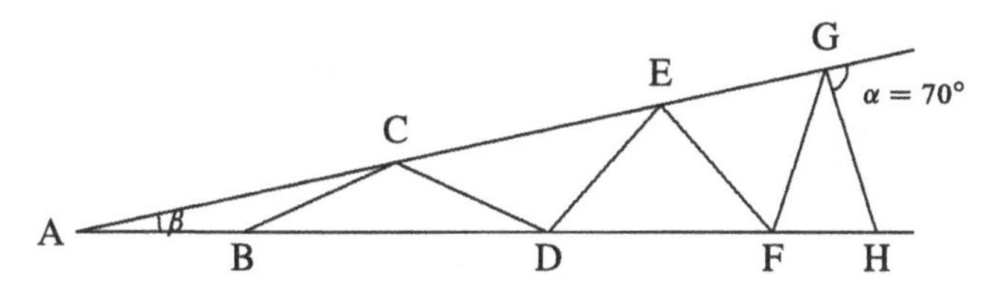

Solution $\angle A = \beta \Rightarrow \angle ACB = \beta \Rightarrow \angle CBD = 2\beta \Rightarrow \angle CDB = 2\beta$

$$\Rightarrow \angle ECD = 3\beta \Rightarrow \angle CED = 3\beta \Rightarrow \angle EDF = 4\beta \Rightarrow \angle EFD = 4\beta$$
$$\Rightarrow \angle GEF = 5\beta \Rightarrow \angle EGF = 5\beta \Rightarrow \angle GFH = 6\beta \Rightarrow \angle GHF = 6\beta$$
$$\Rightarrow \angle\alpha = 7\beta.$$

Therefore $\beta = 10°$.

Example 6. As shown in the diagram, BE and CF bisect $\angle ABD$ and $\angle ACD$ respectively. BE and CF intersect at G. Given that $\angle BDC = 150°$ and $\angle BGC = 100°$, find $\angle A$ in degrees.

Solution Connect BC. Then

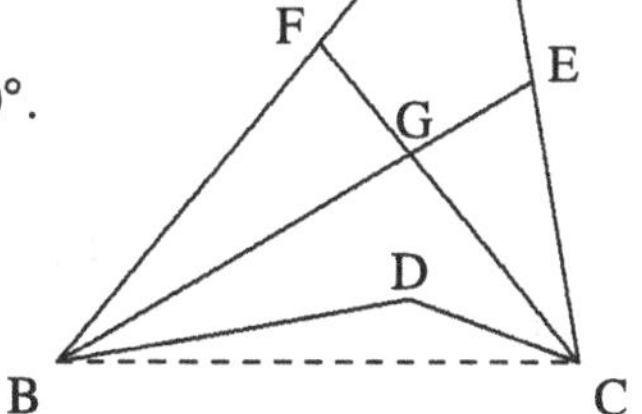

$\because \angle BDC + \angle DBC + \angle DCB = 180°$,
$\therefore \angle DBC + \angle DCB = 180° - 150° = 30°$.
$\because \angle BGC + \angle GBD + \angle GCD + \angle DBC + \angle DCB = 180°$,
$\therefore \angle GBD + \angle GCD = \angle BDC - \angle BGC = 50°$.

Hence

$$\angle ABD + \angle ACD = 2 \cdot 50° = 100°,$$
$$\angle A = 180° - 100° - 30° = 50°.$$

Example 7. (CHINA/1986) As shown in the figure, in $\triangle ABC$, the angle bisectors of the exterior angles of $\angle A$ and $\angle B$ intersect opposite sides at D and E respectively, and $AD = AB = BE$. Then the size of angle A, in degrees, is

(A) 10°, (B) 11°, (C) 12°, (D) None of preceding.

Solution Let

$$\angle A = \angle E = \alpha,$$
$$\angle D = \angle ABD = \beta,$$
$$\angle CBE = \gamma, \angle ACB = \delta.$$

Then $\beta = 2\gamma$ and $\beta = \alpha + \delta$, $\delta = \gamma + \alpha$, so $\beta = 2\alpha + \gamma$. From $2\gamma = \beta = 2\alpha + \gamma$, we obtain $\gamma = 2\alpha$, so $\beta = 4\alpha$.

$$\because \frac{1}{2}(180^\circ - \alpha) + 2\beta = 180^\circ,$$
$$\therefore 4\beta - \alpha = 180^\circ,$$
$$16\alpha - \alpha = 180^\circ,$$
$$\alpha = 12^\circ, \quad \therefore \angle A = 12^\circ.$$

Example 8. There are four points A, B, C, D on the plane, such that any three points are not collinear. Prove that in the triangles ABC, ABD, ACD and BCD there is at least one triangle which has an interior angle not greater than 45°.

Solutions It suffices to discuss the two cases indicated by the following figures:

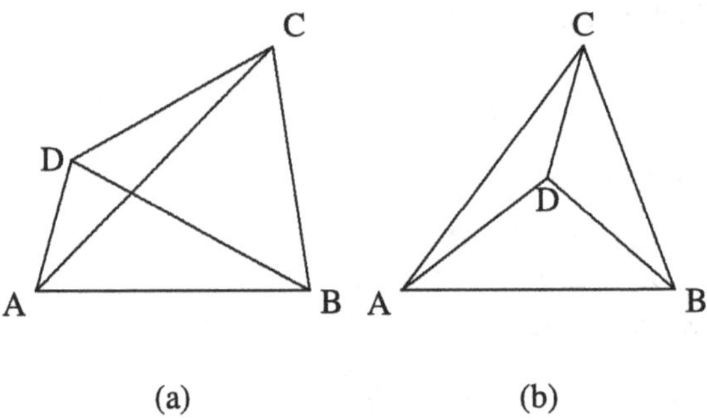

(a) (b)

For case (a), since $\angle DAB + \angle ABC + \angle BCD + \angle CDA = 360^\circ$, at least one of them is not less than 90°. Assuming $\angle CDA \geq 90^\circ$, then in $\triangle CDA$, $\angle DCA + \angle CAD \leq 90^\circ$, so one of them is not greater than 45°.

For case (b), since $\angle ADB + \angle ADC + \angle BDC = 360^\circ$, one of the three angles is greater than 90°, say $\angle ADB > 90^\circ$, then $\angle DAB + \angle DBA < 90^\circ$, so one of $\angle DAB$ and $\angle DBA$ is less than 45°.

Example 9. Given that in a right triangle the length of a leg of the right angle is 11 and the lengths of the other two sides are both positive integers. Find the perimeter of the triangle.

Solution From the given conditions we have

$$n^2 = m^2 + 11^2,$$
$$n^2 - m^2 = 11^2,$$
$$(n-m)(n+m) = 121 = 1 \cdot 121 = 11 \cdot 11,$$

11 n m

therefore

$$n - m = 1, n + m = 121 \quad \text{or} \quad n - m = 11, n + m = 11,$$
$$\therefore n = 61, m = 60. \ (n = 11, m = 0 \text{ is not acceptable.})$$

Thus, the perimeter is $11 + 61 + 60 = 132$.

Testing Questions (A)

1. The sum of all interior angles of a convex n-sided polygon is less than 2007°. Find the maximum value of n.

2. (AHSME/1961) In $\triangle ABC$, $AB = BC$. The points P and Q are on the sides BC and AB respectively, such that $AC = AP = PQ = QB$. then the size of $\angle B$ (in degrees) is

   (A) $25\frac{5}{7}$, (B) $26\frac{1}{3}$, (C) 30, (D) 40, (E) not determined.

3. (CHINA/1997) In a right-angled $\triangle ABC$, $\angle ACB = 90°$, E, F are on AB such that $AE = AC$, $BF = BC$, find $\angle ECF$ in degrees.

4. If the perimeter of a triangle is 17, and the lengths of its three sides are all positive integers, find the number of such triangles.

5. Given that the lengths of three sides, a, b, c of a triangle are positive integers, where $a < b < c$. Find the number of the triangles with $b = 2$.

6. In a right-angled triangle, if the length of a leg is 21, and the lengths of the other two sides are also positive integers, find the minimum value of its possible perimeter.

7. (AHSME/1978) In $\triangle ADE$, $\angle ADE = 140°$. The points B and C are on the sides AD and AE respectively. If $AB = BC = CD = DE$, then $\angle EAD$, in degrees, is

   (A) 5°, (B) 6°, (C) 7.5°, (D) 8°, (E) 10°.

8. (AHSME/1977) In $\triangle ABC$, $AB = AC, \angle A = 80°$. If the points D, E, F are on the sides BC, CA and AB respectively, such that $CE = CD$, $BF = BD$, then $\angle EDF$, in degrees, is

   (A) 30°, (B) 40°, (C)50°, (D) 65°, (E) none of preceding.

9. If the lengths of three sides of a triangle are consecutive positive integers, and its perimeter is less than or equal to 100, how many such acute triangles are there?

10. (AHSME/1996) Triangles ABC and ABD are isosceles with $AB = AC = BD$, and AC intersects BD at E. If AC is perpendicular to BD, then $\angle C + \angle D$ is

    (A) 115°, (B) 120°, (C) 130°, (D) 135°, (E) not uniquely determined.

Testing Questions (B)

1. (CHINA/1991) In $\triangle ABC, \angle A = 70°$, D is on the side AC, and the angle bisector of $\angle A$ intersects BD at H such that $AH : HE = 3 : 1$ and $BH : HD = 5 : 3$. Then $\angle C$ in degrees is

   (A) 45°, (B) 55°, (C) 75°, (D) 80°.

2. (CHINA/1998) In triangle ABC, $\angle A = 96°$. Extend BC to an arbitrary point D. The angle bisectors of angle ABC and $\angle ACD$ intersect at A_1, and the angle bisectors of $\angle A_1BC$ and $\angle A_1CD$ intersect at A_2, and so on. The angle bisectors of $\angle A_4BC$ and $\angle A_4CD$ intersect at A_5. Find the size of $\angle A_5$ in degrees.

3. In $\triangle ABC$, $AB = AC, D, E, F$ are on AB, BC, CA, such that $DE = EF = FD$. Prove that $\angle DEB = \frac{1}{2}(\angle ADF + \angle CFE)$.

4. In right-angled $\triangle ABC$, $\angle C = 90°$, E is on BC such that $AC = BE$. D is on AB such that $DE \perp BC$. Given that $DE + BC = 1$, $BD = \frac{1}{2}$, find $\angle B$ in degrees.

5. (MOSCOW/1952) In $\triangle ABC, AC = BC$, $\angle C = 20°$, M is on the side AC and N is on the side BC, such that $\angle BAN = 50°$, $\angle ABM = 60°$. Find $\angle NMB$ in degrees.

Lecture 10

Pythagoras' Theorem and Its Applications

Theorem I. *(Pythagoras' Theorem) For a right-angled triangle with two legs a, b and hypotenuse c, the sum of squares of legs is equal to the square of its hypotenuse, i.e. $a^2 + b^2 = c^2$.*

Theorem II. *(Inverse Theorem) If the lengths a, b, c of three sides of a triangle have the relation $a^2 + b^2 = c^2$, then the triangle must be a right-angled triangle with two legs a, b and hypotenuse c.*

When investigating a right-angled triangle (or shortly, right triangle), the following conclusions are often used:

Theorem III. *A triangle is a right triangle, if and only if the median on one side is half of the side.*

Theorem IV. *If a right triangle has an interior angle of size $30°$, then its opposite leg is half of the hypotenuse.*

Examples

Example 1. Given that the perimeter of a right angled triangle is $(2 + \sqrt{6})$ cm, the median on the hypotenuse is 1 cm, find the area of the triangle.

Solution The Theorem III implies that $AD = BD = CD = 1$, so $AB = 2$. Let $AC = b$, $BC = a$, as shown in the figure below, then

$$a^2 + b^2 = 2^2 = 4 \quad \text{and} \quad a + b = \sqrt{6}.$$

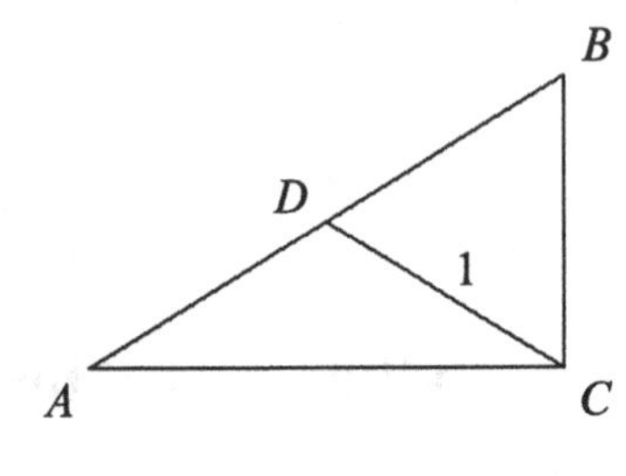

Therefore $6 = (a+b)^2 = a^2 + b^2 + 2ab$, so

$$ab = \frac{6-4}{2} = 1,$$

the area of the triangle ABC is $\frac{1}{2}$.

Example 2. As shown in the figure, $\angle C = 90°$, $\angle 1 = \angle 2$, $CD = 1.5$ cm, $BD = 2.5$ cm. Find AC.

Solution From D introduce $DE \perp AB$, intersecting AB at E. When we fold up the plane that $\triangle CAD$ lies along the line AD, then C coincides with E, so

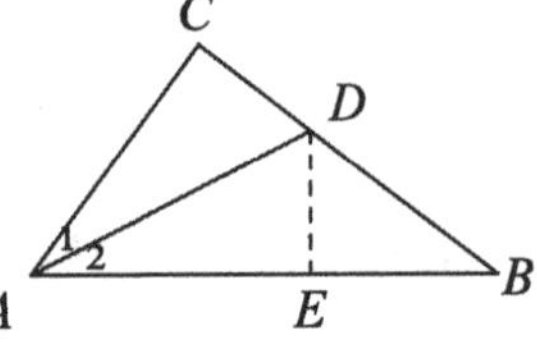

$$AC = AE, \quad DE = CD = 1.5 \text{ (cm)}.$$

By applying Pythagoras' Theorem to $\triangle BED$,

$$BE = \sqrt{BD^2 - DE^2} = \sqrt{6.25 - 2.25} = 2 \text{ (cm)}.$$

Letting $AC = AE = x$ cm and applying Pythagoras' Theorem to $\triangle ABC$ leads the equation

$$(x+2)^2 = x^2 + 4^2,$$
$$4x = 12, \quad \therefore x = 3.$$

Thus $AC = 3$ cm.

Example 3. As shown in the figure, $ABCD$ is a square, P is an inner point such that $PA : PB : PC = 1 : 2 : 3$. Find $\angle APB$ in degrees.

Solution Without loss of generality, we assume that $PA = 1$, $PB = 2$, $PC = 3$. Rotate the $\triangle APB$ around B by 90° in clockwise direction, such that $P \to Q$, $A \to C$, then $\triangle BPQ$ is an isosceles right triangle, therefore

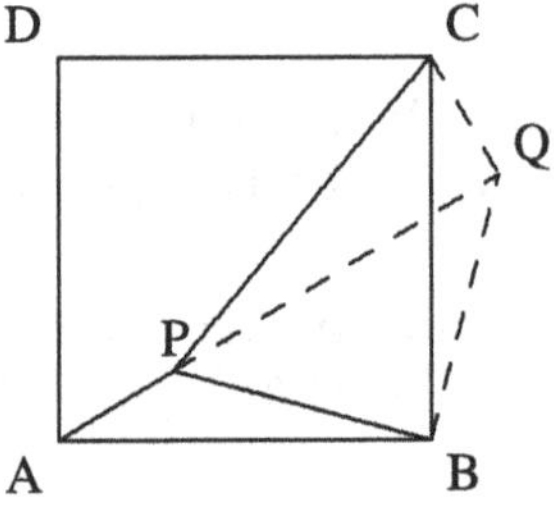

$$PQ^2 = 2PB^2 = 8, CQ^2 = PA^2 = 1,$$

hence, by Pythagoras' Theorem,

$$PC^2 = 9 = CQ^2 + PQ^2, \quad \angle CQP = 90°.$$

Thus, $\angle APB = \angle CQB = 90° + 45° = 135°$.

Example 4. (SSSMO(J)/2003) The diagram shows a hexagon $ABCDEF$ made up of five right-angled isosceles triangles ABO, BCO, CDO, DEO, EFO, and a triangle AOF, where O is the point of intersection of the lines BF and AE. Given that $OA = 8$ cm, find the area of $\triangle AOF$ in cm^2.

Solution From

$$\begin{aligned} OC &= \tfrac{1}{\sqrt{2}}OB = (\tfrac{1}{\sqrt{2}})^2 OA = \tfrac{1}{2}OA, \\ OE &= (\tfrac{1}{\sqrt{2}})^2 OC = \tfrac{1}{4}OA = 2 \text{ (cm)}. \end{aligned}$$

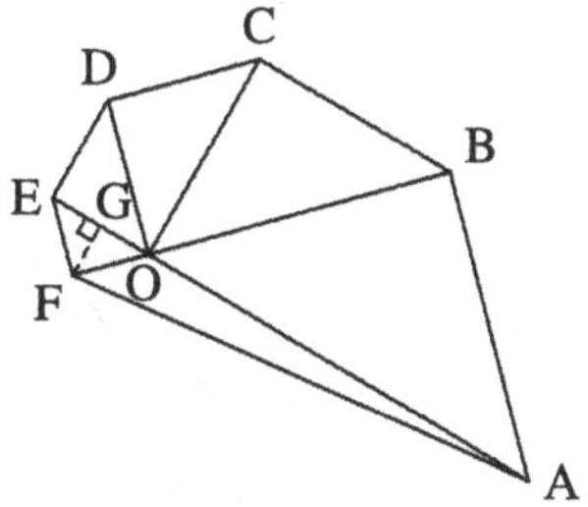

Since Rt$\triangle EFO \sim$ Rt$\triangle ABO$,

$$EF = OF = \frac{1}{4}OB = \frac{1}{4\sqrt{2}}OA.$$

Let $FG \perp AE$ at G, then $FG = \frac{1}{\sqrt{2}}OF$ $= \frac{1}{8}OA = 1$ cm. Thus, the area of $\triangle AOF$, denoted by $[AOF]$, is given by

$$[AOF] = \frac{1}{2}AO \cdot FG = 4 \text{ (cm}^2\text{)}.$$

Example 5. (Formula for median) In $\triangle ABC$, AM is the median on the side BC. Prove that $AB^2 + AC^2 = 2(AM^2 + BM^2)$.

Solution Suppose that $AD \perp BC$ at D. By Pythagoras' Theorem,

$$\begin{aligned} AB^2 &= BD^2 + AD^2 = (BM + MD)^2 + AD^2 \\ &= BM^2 + 2BM \cdot MD + MD^2 + AM^2 - MD^2 \\ &= BM^2 + AM^2 + 2BM \cdot MD. \end{aligned}$$

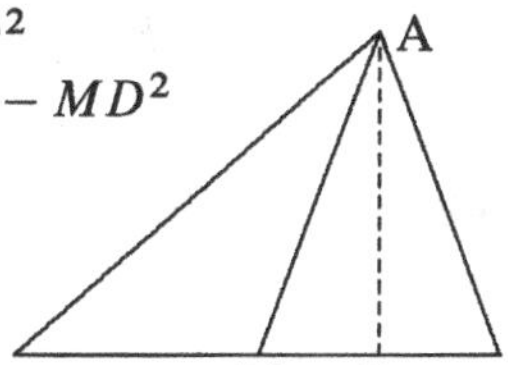

Similarly, we have

$$AC^2 = CM^2 + AM^2 - 2MC \cdot MD.$$

Thus, by adding the two equalities up, since $BM = CM$,

$$AB^2 + AC^2 = 2(AM^2 + BM^2).$$

Note: When AM is extended to E such that $ABEC$ is a parallelogram, then the formula of median is the same as the *parallelogram rule*:

$$AB^2 + BE^2 + EC^2 + CA^2 = AE^2 + BC^2.$$

Example 6. In the figure, $\angle C = 90°$, $\angle A = 30°$, D is the mid-point of AB and $DE \perp AB$, $AE = 4$ cm. Find BC.

Solution Connect BE. Since ED is the perpendicular bisector of AB, $BE = AE$, so $\angle EBD = \angle EBA = \angle A = 30°$, $\angle CBE = 60° - 30° = 30°$, $\therefore CE = \frac{1}{2}BE = DE = \frac{1}{2}AE = 2$ cm.
Now let $BC = x$ cm, then from Pythagoras' Theorem,

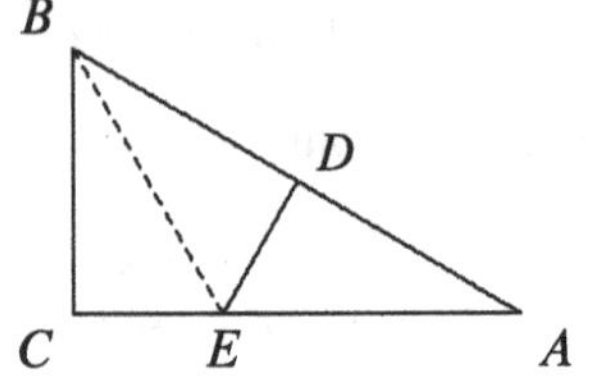

$$(2x)^2 = x^2 + 6^2 \Longrightarrow x^2 = 12$$
$$\Longrightarrow x = \sqrt{12} = 2\sqrt{3} \text{ (cm)}.$$

Thus, $BC = 2\sqrt{3}$ cm.

Example 7. For $\triangle ABC$, O is an inner point, and D, E, F are on BC, CA, AB respectively, such that $OD \perp BC$, $OE \perp CA$, and $OF \perp AB$. Prove that $AF^2 + BD^2 + CE^2 = BF^2 + DC^2 + AE^2$.

Solution By applying the Pythagoras' Theorem to the triangles OAF, OBF, OBD, OCD, OCE and OAE, it follows that

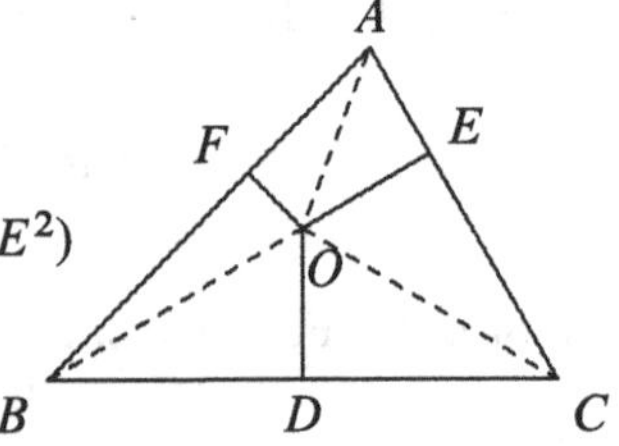

$$\begin{aligned}&AF^2 + BD^2 + CE^2\\ &= AO^2 - OF^2 + BO^2 - OD^2 + CO^2 - OE^2\\ &= (BO^2 - OF^2) + (CO^2 - OD^2) + (AO^2 - OE^2)\\ &= BF^2 + DC^2 + AE^2.\end{aligned}$$

The conclusion is proven.

Example 8. In the diagram given below, P is an interior point of $\triangle ABC$, $PP_1 \perp AB$, $PP_2 \perp BC$, $PP_3 \perp AC$, and $BP_1 = BP_2$, $CP_2 = CP_3$, prove that $AP_1 = AP_3$.

Solution For the quadrilateral AP_1BP, since its two diagonals are perpendicular to each other,

$$\begin{aligned}AP_1^2 + BP^2 &= AF^2 + P_1F^2 + BF^2 + PF^2\\ &= AP^2 + BP_1^2.\end{aligned}$$

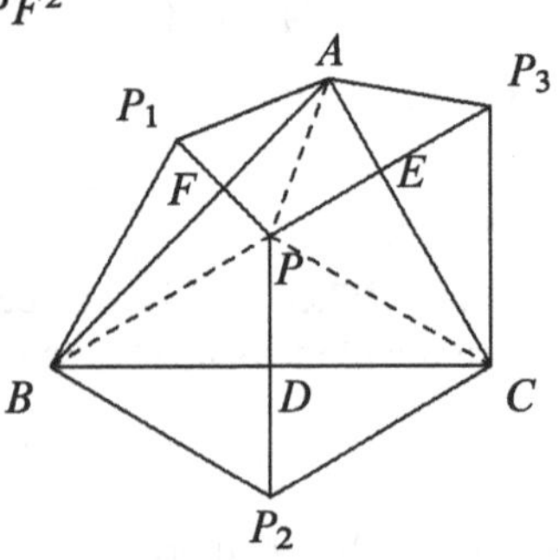

By considering AP_3CP and PCP_2B respectively, it follows similarly that

$$AP^2 + CP_3^2 = AP_3^2 + PC^2,$$
$$BP_2^2 + PC^2 = PB^2 + CP_2^2.$$

Then adding up the three equalities yields

$$AP_1^2 = AP_3^2, \quad \therefore AP_1 = AP_3.$$

Example 9. In square $ABCD$, M is the midpoint of AD and N is the midpoint of MD. Prove that $\angle NBC = 2\angle ABM$.

Solution Let $AB = BC = CD = DA = a$. Let E be the midpoint of CD. Let the lines AD and BE intersect at F.

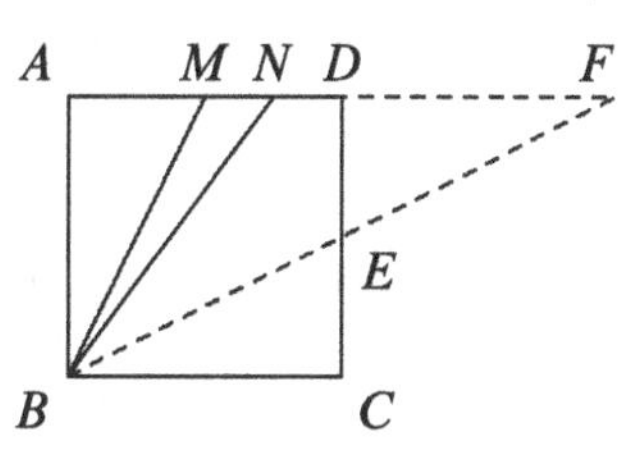

By symmetry, we have $DF = CB = a$. Since right triangles ABM and CBE are symmetric in the line BD, $\angle ABM = \angle CBE$.
It suffices to show $\angle NBE = \angle EBC$, and for this we only need to show $\angle NBF = \angle BFN$ since $\angle DFE = \angle EBC$.

By assumption we have

$$AN = \frac{3}{4}a, \quad \therefore NB = \sqrt{(\frac{3}{4}a)^2 + a^2} = \frac{5}{4}a.$$

On the other hand,

$$NF = \frac{1}{4}a + a = \frac{5}{4}a,$$

so $NF = BN$, hence $\angle NBF = \angle BFN$.

Testing Questions (A)

1. (CHINA/1995) In $\triangle ABC$, $\angle A = 90°$, $AB = AC$, D is a point on BC. Prove that $BD^2 + CD^2 = 2AD^2$.

2. Given that Rt$\triangle ABC$ has a perimeter of 30 cm and an area of 30 cm^2. Find the lengths of its three sides.

3. In the Rt$\triangle ABC$, $\angle C = 90°$, AD is the angle bisector of $\angle A$ which intersects BC at D. Given $AB = 15$ cm, $AC = 9$ cm, $BD : DC = 5 : 3$. Find the distance of D from AB.

4. In the right triangle ABC, $\angle C = 90°$, $BC = 12$ cm, $AC = 6$ cm, the perpendicular bisector of AB intersects AB and BC at D and E respectively. Find CE.

5. In the rectangle $ABCD$, $CE \perp DB$ at E, $BE = \frac{1}{4}BD$ and $CE = 5$ cm. Find the length of AC.

6. In $\triangle ABC$, $\angle C = 90°$, D is the mid-point of AC. Prove that
$$AB^2 + 3BC^2 = 4BD^2.$$

7. In the right triangle ABC, $\angle C = 90°$, E, D are points on AC and BC respectively. Prove that
$$AD^2 + BE^2 = AB^2 + DE^2.$$

8. (CHNMOL/1990) $\triangle ABC$ is an isosceles triangle with $AB = AC = 2$. There are 100 points $P_1, P_2, \ldots, P_{100}$ on the side BC. Write $m_i = AP_i^2 + BP_i \cdot P_iC$ $(i = 1, 2, \ldots, 100)$, find the value of $m_1 + m_2 + \cdots + m_{100}$.

9. In $\triangle ABC$, $\angle C = 90°$, D is the midpoint of AB, E, F are two points on AC and BC respectively, and $DE \perp DF$. Prove that $EF^2 = AE^2 + BF^2$.

10. (CHINA/1996) Given that P is an inner point of the equilateral triangle ABC, such that $PA = 2$, $PB = 2\sqrt{3}$, $PC = 4$. Find the length of the side of $\triangle ABC$.

Testing Questions (B)

1. (SSSMO(J)/2003/Q8) AB is a chord in a circle with center O and radius 52 cm. The point M divides the chord AB such that $AM = 63$ cm and $MB = 33$ cm. Find the length OM in cm.

2. (CHINA/1996) $ABCD$ is a rectangle, P is an inner point of the rectangle such that $PA = 3$, $PB = 4$, $PC = 5$, find PD.

3. Determine whether such a right-angled triangle exists: each side is an integer and one leg is a multiple of the other leg of the right angle.

4. (AHSME/1996) In rectangle $ABCD$, $\angle C$ is trisected by CF and CE, where E is on AB, F is on AD, $BE = 6$ and $AF = 2$. Which of the following is closest to the area of the rectangle $ABCD$?

   (A) 110, (B) 120, (C) 130, (D) 140, (E) 150.

5. (Hungary/1912) Let $ABCD$ be a convex quadrilateral. Prove that $AC \perp BD$ if and only if $AB^2 + CD^2 = AD^2 + BC^2$.

Lecture 11

Congruence of Triangles

Two triangles are called **congruent** if and only if their shapes and sizes are both the same.

In geometry, congruence of triangles is a very important and basic tool in proving the equality relations or inequality relations of two geometric elements (e.g. two segments, two angles, two sums of sides, two differences of angles, etc.). Two triangles are congruent means they are the same in all aspects, so any corresponding geometric elements are equal also.

To prove two geometric elements being equal, it is convenient to construct two congruent triangles such that the two elements are the corresponding elements of the congruent triangles.

To prove two geometric elements are equal or not equal, even though their positions are wide apart, by using the congruence of two triangles, we can move the position of a triangle which carries one element, such that these two elements are positioned together so their comparison becomes much easier.

Basic Criteria for Congruence of Two Triangles

(i) ***S.A.S.***: Two sides and their included angle of one triangle are equal to those in the other triangle correspondingly.

(ii) ***A.A.S.***: Two angles and one side of a triangle are equal to those in the other triangle correspondingly.

(iii) ***S.S.S.***: Three sides of a triangle are equal to those of the other triangle correspondingly.

For right triangles, these criteria can be simplified as follows:

(iv) ***S.A.***: One side and one acute angle of a triangle are equal to those of the other triangle correspondingly.

(v) ***S.S.***: Two sides of a triangle are equal to those of the other triangle correspondingly.

Examples

Example 1. As shown in the diagram, given that in $\triangle ABC$, $AB = AC$, D is on AB and E is on the extension of AC such that $BD = CE$. The segment DE intersects BC at G. Prove that $DG = GE$.

Solution From D introduce $DF \parallel AE$, intersecting BC at F, as shown in the right diagram. Then

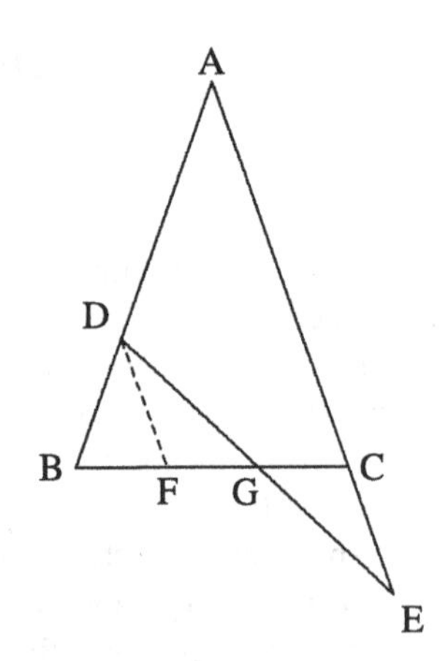

$$\angle FDG = \angle CEG, \angle DGF = \angle EGC.$$

Since $\angle BFD = \angle BCA = \angle DBF$, we have

$$DF = DB = CE.$$

Therefore

$$\triangle DFG \cong \triangle ECG(\text{A.A.S.}),$$

hence $DG = GE$.

Example 2. Given that BE and CF are the altitudes of the $\triangle ABC$. P, Q are on BE and the extension of CF respectively such that $BP = AC, CQ = AB$. Prove that $AP \perp AQ$.

Solution From $AB \perp CQ$ and $BE \perp AC$

$$\angle ABE = \angle QCA.$$

Since $AB = CQ$ and $BP = CA$,

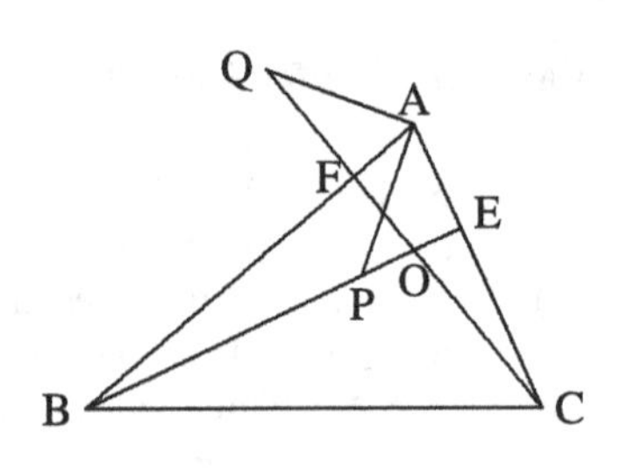

$$\begin{aligned}
&\triangle ABP \cong \triangle QCA \text{ (S.A.S.)},\\
&\therefore \angle BAP = \angle CQA,\\
&\therefore \angle QAP = \angle QAF + \angle BAP\\
&= \angle QAF + \angle CQA\\
&= 180^\circ - 90^\circ = 90^\circ.
\end{aligned}$$

Example 3. (CHINA/1992) In the equilateral $\triangle ABC$, the points D and E are on AC and AB respectively, such that BD and CE intersect at P, and the area of the quadrilateral $ADPE$ is equal to area of $\triangle BPC$. Find $\angle BPE$.

Solution From E and D introduce $EF \perp AC$ at F and $DG \perp BC$ at G.

The condition $[ADPE] = [BPC]$ implies that

$$[ACE] = [CBD].$$

Since $AC = BC$, so $EF = DG$. Since $\angle A = \angle C$, so Rt$\triangle AEF \cong$ Rt$\triangle CDG$ (A.S.). therefore $AE = CD$, hence

$$\triangle AEC \cong \triangle CDB\text{(S.A.S.)}.$$

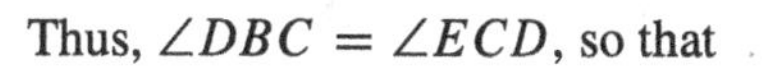

Thus, $\angle DBC = \angle ECD$, so that

$$\angle BPE = \angle PBC + \angle PCB = \angle PCD + \angle PCB = 60°.$$

Example 4. (CHINA/1991) Given that ABC is an equilateral triangle of side 1, $\triangle BDC$ is isosceles with $DB = DC$ and $\angle BDC = 120°$. If points M and N are on AB and AC respectively such that $\angle MDN = 60°$, find the perimeter of $\triangle AMN$.

Solution $\because \angle DBC = \angle DCB = 30°, \therefore DC \perp AC, DB \perp AB.$

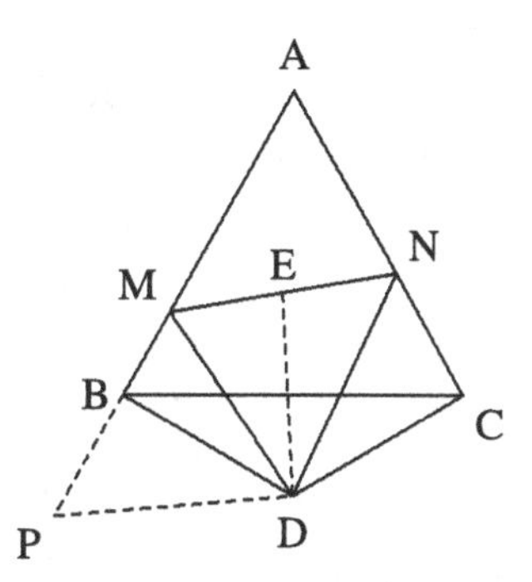

Extending AB to P such that $BP = NC$, then $\triangle DCN \cong \triangle DBP$ (S.S.), therefore $DP = DN$. $\angle PDM = 60° = \angle MDN$ implies that

$$\triangle PDM \cong \triangle MDN, \text{ (S.A.S.)},$$
$$\therefore PM = MN,$$
$$\therefore MN = PM = BM + PM = BM + NC.$$

Thus, the perimeter of $\triangle AMN$ is 2.

Note: Here the congruence $\triangle PDM \cong \triangle MDN$ is obtained by rotating $\triangle DCN$ to the position of $\triangle DBP$ essentially .

Example 5. As shown in the figure, in $\triangle ABC$, D is the mid-point of BC, $\angle EDF = 90°$, DE intersects AB at E and DF intersects AC at F. Prove that

$$BE + CF > EF.$$

Solution In this problem, for the comparison of $BE + CF$ and EF it is needed to move the segments BE, EF, CF together in a same triangle, and constructing congruent triangles can complete this task as follows.

Rotate $\triangle DCF$ around D by 180° in clockwise direction, then

$$C \to B, \quad F \to G.$$

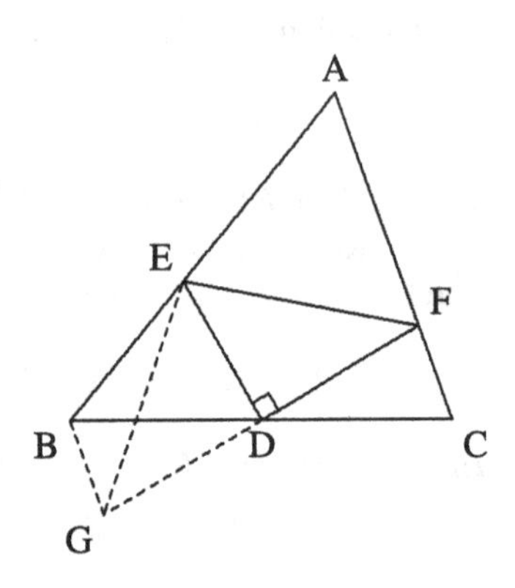

Connect BG, EG, GD. Since $ED \perp GF$ and $GD = DF$, we have

$$\triangle EDG \cong \triangle EDF, \text{ (S.S.)},$$

hence

$$EF = EG < BE + BG = BE + CF.$$

Example 6. (CHINA/1999) Given that $\triangle ABC$ is a right-angled isosceles triangle with $\angle ACB = 90°$. D is the mid-point of BC, CE is perpendicular to AD, intersecting AB and AD at E and F respectively. Prove that $\angle CDF = \angle BDE$.

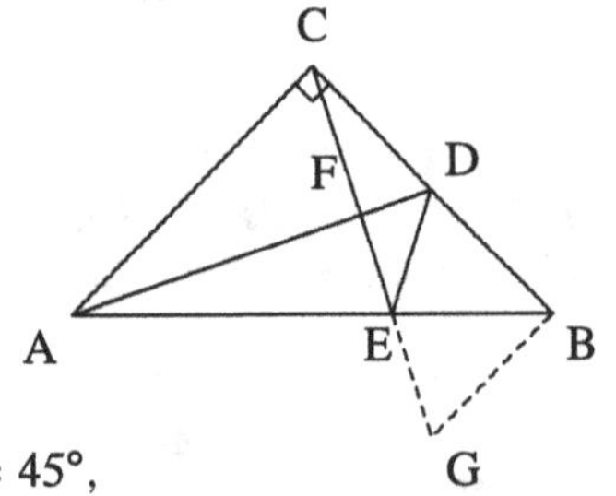

Solution It's inconvenient to compare $\angle CDF$ and $\angle BDE$ directly. To change the position of $\angle CDF$, suppose that the perpendicular line from B to BC intersects the line CE at G. Since $AC = CB$,

$$\angle CAD = \angle BCG = 90° - \angle ACF,$$
$$\triangle ACD \cong \triangle CBG \text{ (S.A.)},$$
$$\therefore \angle CDF = \angle BGC = \angle BGE.$$
$$\because BD = CD = BG \text{ and } \angle DBE = \angle GBE = 45°,$$

$$\therefore \triangle BGE \cong \triangle BDE \text{ (S.A.S.), hence } \therefore \angle CDF = \angle BGE = \angle BDE.$$

Example 7. (CHINA/1992) In the square $ABCD$, E is the midpoint of AD, BD and CE intersect at F. Prove that $AF \perp BE$.

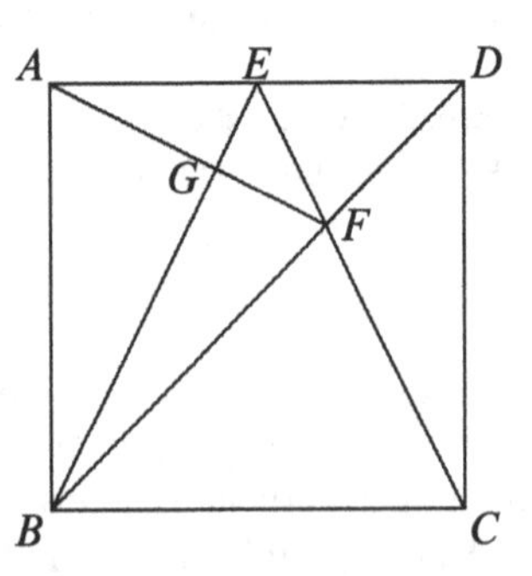

Solution Let G be the point of intersection of AF and BE. It suffices to show

$$\angle EAG = \angle ABG.$$

By symmetry we have

$$\triangle ABE \cong \triangle DCE, \triangle ADF \cong \triangle CDF,$$

therefore $\angle EAG = \angle DCF = \angle ABG$.

Example 8. (CHINA/1992, 1993) In the graph, triangles ABD and BEC are both equilateral with A, B, C being collinear, M and N are midpoints of AE and CD respectively, AE intersects BD at G and CD intersects BE at H. Prove that (i) $\triangle MBN$ is equilateral, (ii) $GH \parallel AC$.

Solution (i) From $AB = BD, BE = BC, \angle ABE = \angle DBC = 120°$

$$\triangle ABE \cong \triangle DBC \text{ (S.A.S.)},$$
$$\therefore \angle MAB = \angle NDB, MA = ND,$$

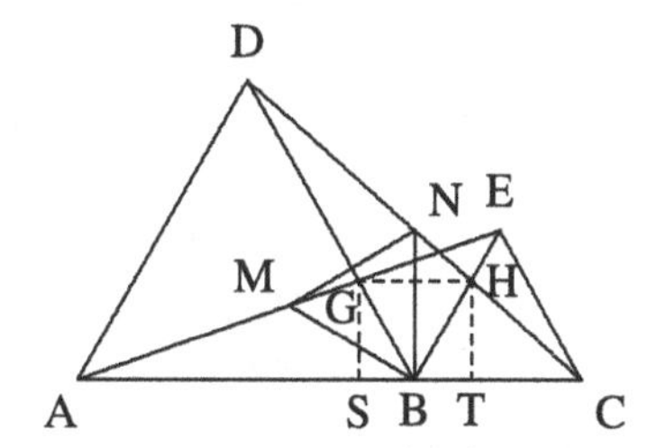

which implies $\triangle MAB \cong \triangle NDB$ (S.A.S.).

$$\therefore MB = NB \text{ and } \angle ABM = \angle DBN$$
$$\angle MBN = \angle MBD + \angle ABM = \angle ABD$$
$$= 60°, \implies \triangle MBN \text{ is equilateral.}$$

(ii) From G, H introduce $GS \perp AC$ at S and $HT \perp AC$ at T respectively. Then

$$\angle GBA = \angle ECA = \angle HBC = \angle DAC = 60° \implies GB \parallel CE, \quad HB \parallel AD,$$
$$\implies GB = \frac{AB}{AC} \cdot CE, HB = \frac{BC}{AC} \cdot AD \implies GB = HB.$$

Since $\angle GBS = \angle HBT = 60°$, so Rt$\triangle GBS \cong$ Rt$\triangle HBT$ (S.A.). Thus, $GS = HT$, i.e. $GH \parallel AC$.

Testing Questions (A)

1. In $\triangle ABC$, $\angle ACB = 60°$, $\angle BAC = 75°$, $AD \perp BC$ at D, $BE \perp AC$ at E, AD intersects BE at H, Find $\angle CHD$ in degrees.

2. $\triangle ABC$ is equilateral, D is an inner point of $\triangle ABC$ and P is a point outside $\triangle ABC$ such that $AD = BD$, $AB = BP$, and BD bisects $\angle CBP$. Find $\angle BPD$.

3. Given that the side of the square $ABCD$ is 1, points P and Q are on AB and AD respectively, such that the perimeter of $\triangle APQ$ is 2. Find $\angle PCQ$ in degrees by use of congruence of triangles.

4. $ABCD$ is a square, E and F are the midpoints of the sides AB and BC respectively. If M is the point of intersection of CE and DF, prove that $AM = AD$.

5. $ABCD$ is a trapezium with $AD \parallel BC$, $\angle ABC = \angle BAD = 90°$, and $DE = EC = BC$. Prove that $\angle DAE = \frac{1}{3}\angle AEC$.

6. (MOSCOW/1952) In an isosceles triangle ABC, $AB = BC$, $\angle B = 20°$. M, N are on AB and BC respectively such that $\angle MCA = 60°$, $\angle NAC = 50°$. Find $\angle NMC$ in degrees.

7. Given that $\triangle ABC$ is an isosceles right triangle with $AC = BC$ and $\angle ACB = 90°$. D is a point on AC and E is on the extension of BD such that $AE \perp BE$. If $AE = \frac{1}{2}BD$, prove that BD bisects $\angle ABC$.

8. (CHINA/1999) In the square $ABCD$, $AB = 8$, Q is the midpoint of the side CD. Let $\angle DAQ = \alpha$. On CD take a point P such that $\angle BAP = 2\alpha$. If $AP = 10$, find CP.

9. (CHINA/1992) In the pentagon $ABCDE$, $\angle ABC = \angle AED = 90°$, $AB = CD = AE = BC + DE = 1$. Find the area of $ABCDE$.

10. (NORTH EUROPE/2003) D is an inner point of an equilateral $\triangle ABC$ satisfying $\angle ADC = 150°$. Prove that the triangle formed by taking the segments AD, BD, CD as its three sides is a right triangle.

Testing Questions (B)

1. (CHINA/1996) Given that the segment BD is on a line ℓ. On one side of ℓ take a point C and construct two squares $ABCK$ and $CDEF$ respectively outside the $\triangle CBD$. Let M be the midpoint of the segment AE, prove that the position of M is independent of the choice of the position of C.

2. (CHINA/1998) In Rt$\triangle ABC$, $\angle C = 90°$, $CD \perp AB$ at D, AF bisects $\angle A$, intersecting CD and CB at E and F respectively. If EG is parallel to AB, intersecting CB at G, prove that $CF = GB$.

3. (CHINA/1994) In $\triangle ABC$, $AC = 2AB$ and $\angle A = 2\angle C$. Prove that $AB \perp BC$.

4. (CHINA/2000) In a given quadrilateral $ABCD$, $AB = AD, \angle BAD = 60°, \angle BCD = 120°$. Prove that $BC + DC = AC$.

5. In $\triangle ABC$, $\angle ABC = \angle ACB = 80°$. The point P is on AB such that $\angle BPC = 30°$. Prove that $AP = BC$.

Lecture 12

Applications of Midpoint Theorems

In a triangle, the segment joining midpoints of two sides is called a **midline** of the triangle. A triangle has three midlines.

In a trapezium, the segment joining the midpoints of two legs is called the **midline** of the trapezium.

Theorem I. *For any triangle ABC, if D and E are on AB and AC respectively, then $DE \parallel BC$ and $DE = \frac{1}{2}BC$ if and only if D, E are midpoints of AB and AC respectively.*

Theorem II. *For a trapezium $ABCD$ with $AB \parallel CD$, if E, F are the midpoints of AD and BC respectively, then $EF \parallel AB \parallel CD$, and $EF = \frac{1}{2}(AB + CD)$.*

In geometry, these two theorems are often used, since the endpoints of a midline are midpoints of sides, so many problems mentioning determining midpoints can be solved by using midlines.

Since a midline is half of the third side for triangles, or half of the sum of two bases for trapezia, the midlines can be taken as a tool to change the segments to be compared or identified to half or double of these segments, so that their comparison becomes much easier.

Examples

Example 1. In the figure, D, E are points on AB and AC such that $AD = DB$, $AE = 2EC$, and BE intersects CD at point F. Prove that $4EF = BE$.

Solution It is difficult to compare the lengths of EF and BE since they are on a same line. Here we can use a midline as a ruler to measure them.

Let M be the midpoint of AE. Connect DM. By applying the midpoint theorem to $\triangle ABE$ and $\triangle CDM$ respectively, it follows that

$$DM = \frac{1}{2}BE,$$
$$EF = \frac{1}{2}DM$$
$$\therefore EF = \frac{1}{4}BE, \quad \text{i.e. } BE = 4EF.$$

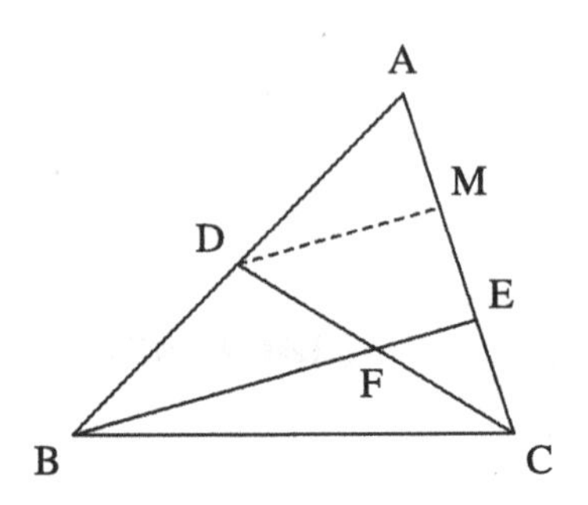

Example 2. Given that $ABCD$ is a convex quadrilateral, $\angle ABC = \angle CDA = 90°$, and $\angle BCD > \angle BAD$, as shown in the diagram below. Prove that $AC > BD$.

Solution Extend AB, AD to E, F respectively, such that $AB = BE$ and $AD = DF$. Then, by the midpoint theorem,

$$BD \parallel EF \quad \text{and} \quad EF = 2BD.$$

Since BC, DC are the perpendicular bisector of AE, AF respectively,

$$EC = AC = FC.$$

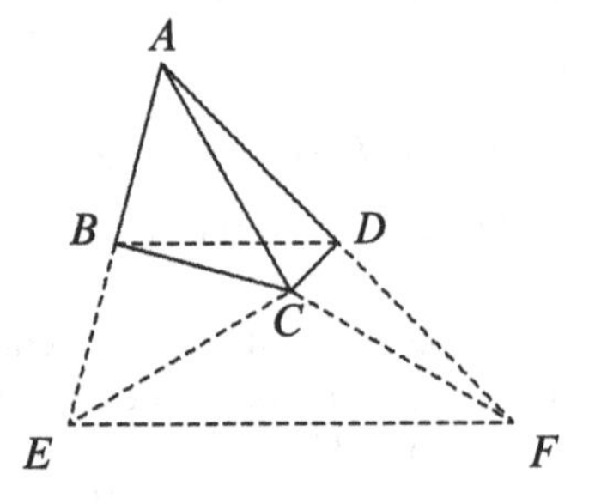

From the triangle inequality,

$$EC + FC > EF, \quad \text{i.e.}$$
$$2AC > 2BD, \quad \therefore AC > BD.$$

Example 3. As shown in the figure, in $\triangle ABC$, $\angle B = 2\angle C$, AD is perpendicular to BC at D and E is the midpoint of BC. Prove that $AB = 2DE$.

Solution Let F be the midpoint of AC, connect EF, DF. By the midpoint theorem, $AB = 2EF$, it suffices to show $DE = EF$. Since DF is the median on the hypotenuse AC of the right triangle ADC, $DF = FC = AF$, so $\angle CDF = \angle C$. Since $EF \parallel AB$,

$$\angle CEF = \angle B = 2\angle C,$$
$$\therefore \angle DFE = \angle CEF - \angle CDF = \angle C$$
$$= \angle CDF, \quad \text{hence } DE = EF.$$

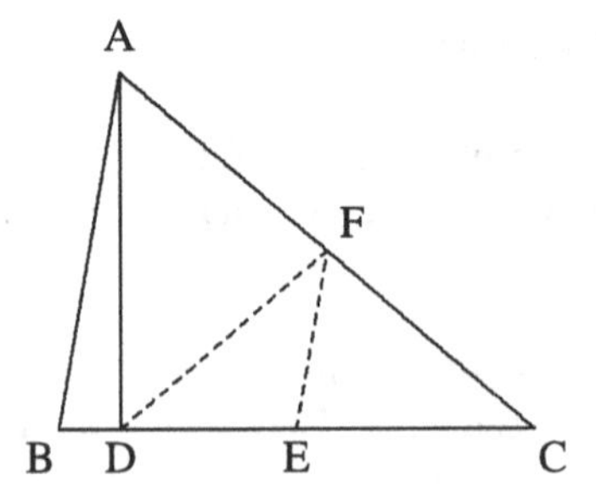

Example 4. In the figure, $AB = CD$, E, F are the midpoints of AD and BC respectively. Let BA intersect FE at M. Prove that $AM = AE$.

Solution The given condition $AB = CD$ and the goal $AM = AE$ have no direct relation. Here the midpoint theorem is the bridge to connect them.

Connect BD. Let P be the midpoint of BD, connect PE, PF. Then, by the midpoint theorem,

$$PE = \frac{1}{2}AB = \frac{1}{2}CD = PF$$

and $PE \parallel BM, AC \parallel PF$. Therefore

$$\angle AME = \angle PEF = \angle PFE = \angle AEM,$$

$$\therefore AM = AE.$$

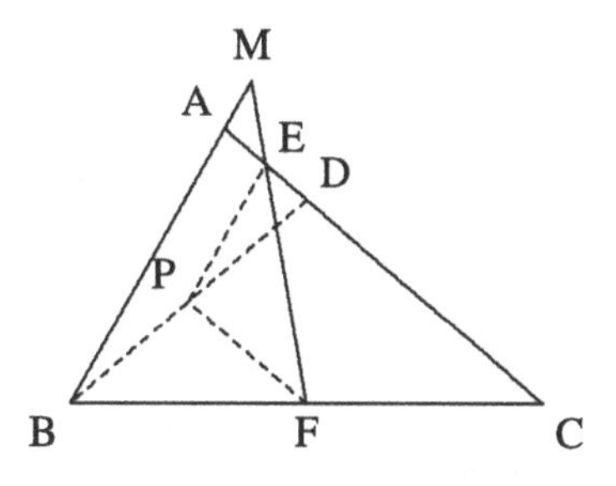

Example 5. For a quadrilateral $ABCD$, E, F are the midpoints of AB and BC respectively, DE and DF intersect the diagonal AC at M and N respectively, such that $AM = MN = NC$. Prove that $ABCD$ is a parallelogram.

Solution We first show that $MBND$ is a parallelogram, then show $ABCD$ is a parallelogram. At the first step, the midpoint theorem plays an essential role.

Connect BM, BD, BN. $\because AE = BE$, $BF = FC$, and $AM = MN = NC$, $EM \parallel BN$ and $FN \parallel BM$, $\therefore MBND$ is a parallelogram. Then

$$BM = ND \text{ and}$$
$$\angle AMB = \angle FNM = \angle CND,$$
$$\triangle AMB \cong \triangle CND \text{ (S.A.S.)},$$
$$\therefore AB = CD, \quad \angle BAC = \angle DCN,$$

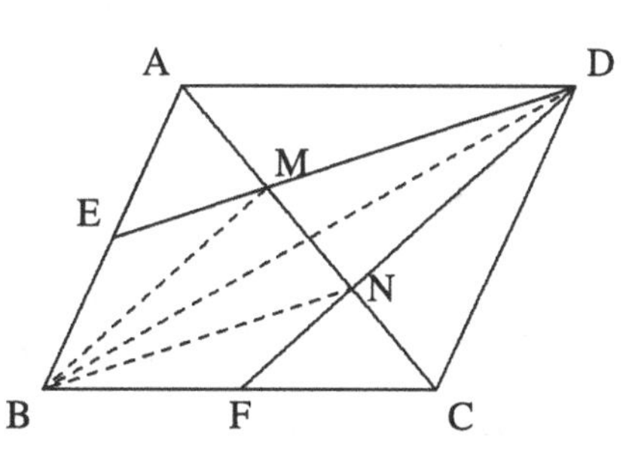

so $AB \parallel CD$ also. Thus, $ABCD$ is a parallelogram.

Example 6. (CHNMOL/1997) In the trapezium $ABCD$, $AD \parallel BC, \angle B = 30°$, $\angle C = 60°$, E, M, F, N are the midpoints of AB, BC, CD, DA respectively. Given that $BC = 7, MN = 3$. Find EF.

Solution By applying the midpoint theorem to the trapezium, then $EF = \frac{1}{2}(AD + BC)$, so it is important to find AD.

Through N we introduce $NG \parallel AB, NH \parallel CD$, intersecting BC at G, H respectively. Since $AD \parallel BC$, the quadrilaterals $ABGN$ and $NHCD$ are both parallelograms.

$$\therefore BG = AN = CH = ND \text{ and}$$
$$AB \parallel NG, CD \parallel NH.$$
$$\because \angle NGH = \angle ABH = 30°, \text{ and}$$
$$\angle NHG = \angle DCG = 60°,$$
$$\angle GNH = 180° - 30° - 60° = 90°.$$
$$\because BM = CM \Longrightarrow GM = HM,$$
$$\therefore GH = 2NM = 6, AD = 7 - 6 = 1.$$
$$\therefore EF = \tfrac{1}{2}(1 + 7) = 4.$$

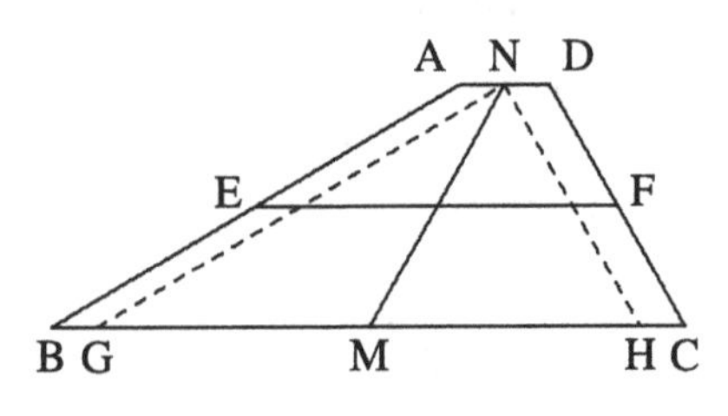

Example 7. (CHINA/1997) In the trapezium $ABCD$, $AB \parallel CD, \angle DAB = \angle ADC = 90°$, and the $\triangle ABC$ is equilateral. Given that the midline of the trapezium $EF = 0.75a$, find the length of the lower base AB in terms of a.

Solution From the given conditions,

$$\angle DAC = 30°, \ \therefore CD = \frac{1}{2}AC = \frac{1}{2}AB.$$

By the midpoint theorem,

$$EF = \tfrac{1}{2}(CD + AB) = \tfrac{3}{4}AB,$$
$$\therefore AB = a.$$

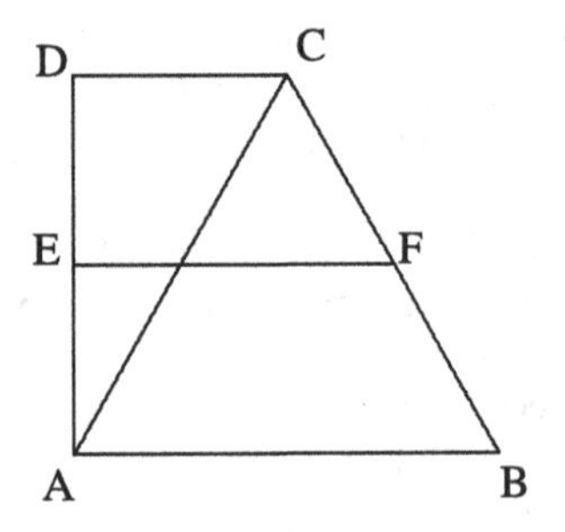

Example 8. (MOSCOW/1995) In a given convex quadrilateral $ABCD$, O is an inner point of $ABCD$ such that $\angle AOB = \angle COD = 120°$, $AO = OB$, $CO = OD$. Given that K, L, M are the midpoints of the segments AB, BC, CD respectively, prove that $\triangle KLM$is equilateral.

Solution It suffices to show that $KL = ML$ and $\angle KLM = 60°$. The conclusion can be obtained by the midpoint theorem.
Let N, P be the midpoints of OB, OC respectively. Connect NK, NL, PL, PM. Then

$$KN = \tfrac{1}{2}OA = \tfrac{1}{2}OB = PL,$$
$$NL = \tfrac{1}{2}OC = \tfrac{1}{2}OD = PM.$$

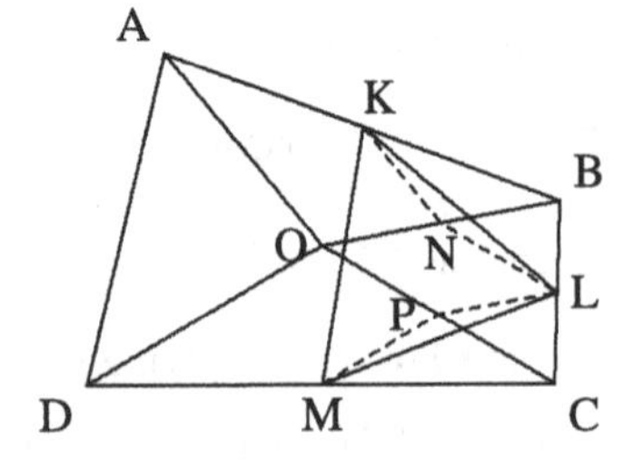

$$\because NK \parallel OA,\ NL \parallel OC,\ PL \parallel OB,\ PM \parallel OD \text{ and}$$
$$\angle KNL = \angle AOC = 120° + \angle BOC = \angle BOD = \angle LPM,$$
$$\therefore \triangle KNL \cong \triangle LPM \text{ (S.A.S.), hence } KL = LM.$$

On the other hand, we have $\angle KLM = \angle KLN + \angle NLP + \angle PLM = \angle PML + \angle LPC + \angle PLM = 180° - \angle CPM = 180° - 120° = 60°$, therefore $\triangle KLM$ is equilateral.

Example 9. (CHNMO/TST/1995) Given that the points P and Q are on the sides AB and AC of an acute triangle ABC respectively. D is an inner point of $\triangle ABC$ such that $PD \perp AB$ at P and $QD \perp AC$ at Q.If M is the midpoint of the side BC, prove that $PM = QM$ if and only if $\angle BDP = \angle CDQ$.

Solution *Sufficiency*: Suppose $\angle BDP = \angle CDQ$. Let E, F be the midpoints of BD, CD respectively. Connect EP, ME, MF, FQ. Then

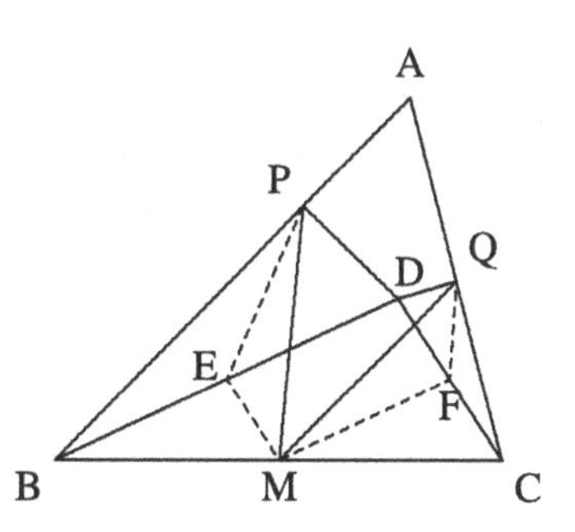

$$EP = \tfrac{1}{2}BD = MF,\ ME = \tfrac{1}{2}CD = FQ,$$
$$\because \angle BDP = \angle CDQ \Longrightarrow \angle PBD = \angle QCD,$$
$$\therefore \angle PED = 2\angle PBD = 2\angle DCQ = \angle DFQ,$$
since $DEMF$ is a parallelogram, therefore $\angle DEM = \angle DFM$, so $\angle PEM = \angle MFQ$,

thus $\triangle PEM \cong \triangle MFQ$ (S.A.S.), hence $PM = QM$.

Necessity: If $PM = QM$, then $\triangle PEM \cong \triangle MFQ$ (S.S.S.), so $\angle PEM = \angle MFQ$, $\angle DEM = \angle MFD$ (by the same reason as above), therefore $\angle PED = \angle DFQ$, i.e. $2\angle PBE = 2\angle DCQ$ or $\angle PBE = \angle DCQ$. Thus, $\angle BDP = 90° - \angle PBD = 90° - \angle DCQ = \angle CDQ$.

Testing Questions (A)

1. Given that $ABCD$ is a quadrilateral, E and F are midpoints of the sides AD and BC of $ABCD$. Suppose that $AB \nparallel CD$. Prove that $EF < \frac{1}{2}(AB + CD)$.

2. In a trapezium $ABCD$, $AB \parallel CD$ and $AB = 2CD$. M, N are the midpoints of the diagonals AC and BD respectively. Let the perimeter of $ABCD$ be

l_1, the perimeter of the quadrilateral $CDMN$ be l_2 and $l_1 = nl_2$, find the value of n.

3. In a square $ABCD$, let O be the intersection point of the diagonals AC and BD. Let the angle bisector of $\angle CAB$ intersect BD at E and BC at F. Prove that $2OE = CF$.

4. In $\triangle ABC$, let E be the midpoint of BC and let D be the foot of the altitude from A to BC. Suppose $AB = 2DE$. Prove that $\angle B = 2\angle C$.

5. $ABCD$ is a trapezium such that $AB \parallel DC, AD = BC$. Given that AC intersect BD at O, P, Q, R are the midpoints of AO, DO and BC respectively, and $\angle AOB = 60°$. Prove that $\triangle PQR$ is equilateral.

6. In the $\triangle ABC$, BE is the angle bisector of the $\angle ABC$, AD is the median on the side BC, and AD intersects BE at O perpendicularly. Given $BE = AD = 4$, find the lengths of three sides of $\triangle ABC$.

7. ABC is a given triangle. If the sides AB and AC are taken as hypotenuses of two right triangles ABD and ACE outside the $\triangle ABC$, respectively, such that $\angle ABD = \angle ACE$. Prove that $DM = EM$, where M is the midpoint of BC.

Testing Questions (B)

1. (CHINA/1999) In a quadrilateral $ABCD$, $AD > BC$, E and F are the midpoints of AB and CD respectively. Suppose that the lines AD and BC intersect FE produced at H and G respectively. Prove that $\angle AHE < \angle BGE$.

2. In $\triangle ABC$, let BC be produced to a point M. Let D, E, N be the midpoints of AB, AC and BM respectively. Let H be the midpoint of EN. Join DH and extend DH to meet BM at F. Prove that $CF = FM$.

3. In the right triangle ABC, $\angle ABC = 90°$, $AB = BC$. Let D and E be points on AB and BC respectively such that $AD = CE$. Let M and N be points on AC such that DM and BN are perpendicular to AE. Prove that $MN = NC$.

4. (CHINA/1997) $ABCD$ is a quadrilateral with $AD \parallel BC$. If the angle bisector of $\angle DAB$ intersects CD at E, and BE bisects $\angle ABC$, prove that $AB = AD + BC$.

Lecture 13

Similarity of Triangles

Two triangles are called **similar** if we can get two congruent triangles after enlarging or compressing the sides of one of them according to an equal ratio. That is, two triangles are similar means they have a same shape but may have different sizes.

Criteria for Similarity of Two Triangles

(I) Each pair of corresponding angles are equal (A.A.A.);
(II) All corresponding sides are proportional (S.S.S.);
(III) Two pairs of corresponding sides are proportional, and the included corresponding angles are equal (S.A.S.);
(IV) For two right triangles, a pair of two corresponding acute angles are equal (A.A.);
(V) Among the three pairs of corresponding sides two pairs are proportional (S.S.).

Basic Properties of Two Similar Triangles

(I) For two similar triangles, their corresponding sides, corresponding heights, corresponding medians, corresponding angle bisectors, corresponding perimeters are all proportional with the same ratio;
(II) Consider the similarity as a transformation from one triangle to other, then this transformation keeps many features of a graph unchanged: each interior angle is unchanged; any two parallel lines are still parallel, the angle formed by two intersected lines keeps unchanged, and collinear points remained collinear.
(III) For two similar triangles, the ratio of their areas is equal to square of the ratio of their corresponding sides.

Important Proportional Properties of Segments

When by $a, b, c, \ldots$ we denote the lengths of segments, the following proportional properties hold, which are the same as in algebra:

(1) $\frac{a}{b} = \frac{c}{d} \Longrightarrow ad = bc;$

(2) $\frac{a}{b} = \frac{c}{d} \Longrightarrow \frac{a+b}{b} = \frac{c+d}{d};$

(3) $\frac{a}{b} = \frac{c}{d} \Longrightarrow \frac{a-b}{b} = \frac{c-d}{d};$

(4) $\frac{a}{b} = \frac{c}{d} \Longrightarrow \frac{a+b}{a-b} = \frac{c+d}{c-d}$ if $a - b \neq 0$ or $c - d \neq 0$;

(5) $\frac{a}{b} = \frac{c}{d} = \cdots = \frac{m}{n}$ $(bd \cdots n \neq 0)$ $\Longrightarrow$ $\frac{a+c+\cdots+m}{b+d+\cdots+n} = \frac{a}{b} = \frac{c}{d} = \cdots = \frac{m}{n}$ if $b + d + \cdots + n \neq 0$.

Examples

Example 1. Prove the following:

(1) When two straight lines are cut by three parallel lines, the two segments between two adjacent parallel lines are proportional.

(2) When a straight line which is parallel to one side of a triangle cuts the other sides of the triangle at two points, the three sides of the derived triangle must be proportional to the three sides of the original triangle, correspondingly.

Solution (1) If the two lines ℓ_1 and ℓ_2 are parallel, the conclusion is obvious. If $\ell_1 \nparallel \ell_2$, suppose that O is their point of intersection. If three parallel lines ℓ_3, ℓ_4, ℓ_5 intersect ℓ_1 and ℓ_2 at A_1, B_1, C_1 and A_2, B_2, C_2 correspondingly, as shown in the right diagram, let

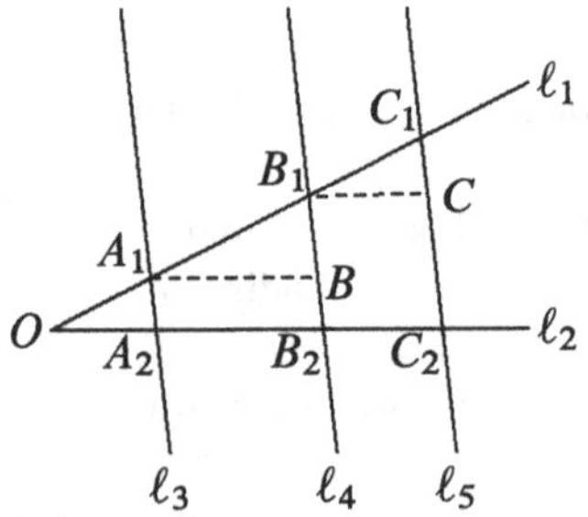

$A_1B \parallel \ell_2$, intersecting ℓ_4 at B,
$B_1C \parallel \ell_2$, intersecting ℓ_5 at C,

then $A_1A_2B_2B$ and $B_1B_2C_2C$ are parallelograms,

$$\therefore A_1B = A_2B_2, \quad B_1C = B_2C_2. \quad \because \triangle A_1BB_1 \sim \triangle B_1CC_1 \text{ (A.A.A.)},$$

$$\therefore \frac{A_1B_1}{B_1C_1} = \frac{A_1B}{B_1C}. \quad \text{Thus } \frac{A_1B_1}{B_1C_1} = \frac{A_2B_2}{B_2C_2}.$$

(2) Given $\triangle ABC$. If the line $\ell \parallel BC$, intersecting AB, AC at B_1, C_1 respectively. Then

$$\angle AB_1C_1 = \angle ABC, \quad \angle AC_1B_1 = \angle ACB,$$
$$\therefore \triangle AB_1C_1 \sim \triangle ABC \text{ (A.A.A.)},$$

therefore the sides of the two triangles are proportional correspondingly.

Example 2. (Angle Bisector Theorem) For any triangle, the angle bisector of any interior angle must cut the opposite side into two segments, such that their ratio is equal to the ratio of the two sides of the angle, correspondingly.

Solution In $\triangle ABC$, Let AD be the angle bisector of $\angle BAC$, where AD and BC intersect at D.

From C introduce $CE \parallel AD$, intersecting the extension of BA at E. Then

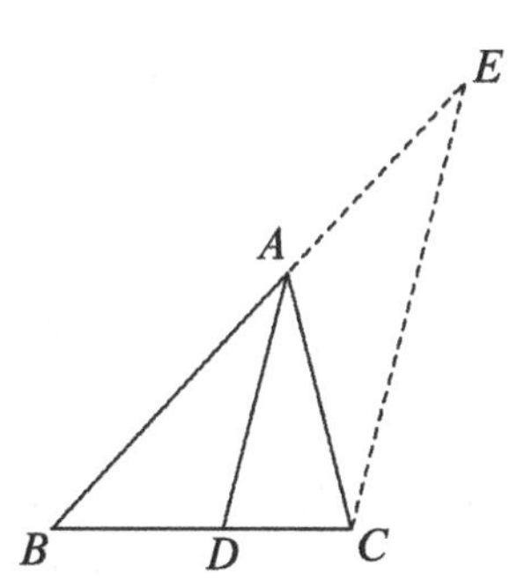

$$\angle ECA = \angle DAC = \angle BAD$$
$$= \angle AEC, \therefore AC = AE.$$
$$\because AD \parallel EC,$$
$$\therefore \frac{AB}{AE} = \frac{BD}{CD},$$
$$\therefore \frac{AB}{AC} = \frac{BD}{CD}.$$

Note: When AD is the angle bisector of the exterior angle of angle BAC, similarly, it is also true that

$$\frac{AB}{AC} = \frac{BD}{CD}.$$

Example 3. (Projection Theorem of Right Triangles) In the right triangle ABC, $\angle ACB = 90°$. Then

$$CD^2 = AD \cdot DB, \qquad AC^2 = AD \cdot AB, \qquad BC^2 = BD \cdot BA.$$

Solution From $\angle ACD = 90° - \angle A = \angle CBD$, therefore Rt$\triangle ACD \sim$ Rt$\triangle CBD$ (A.A.A.), we have

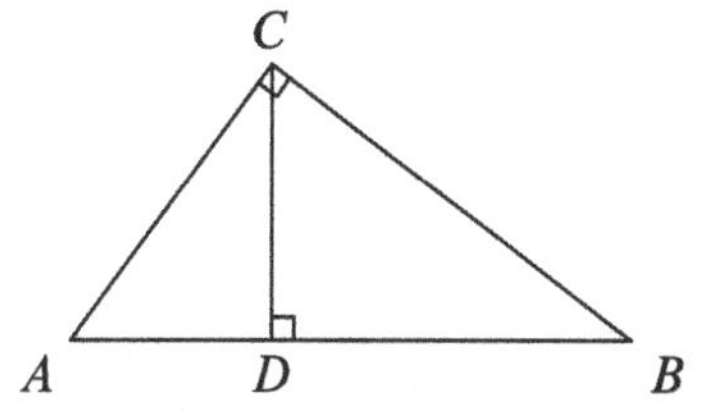

$$\frac{CD}{AD} = \frac{BD}{CD},$$
i.e. $CD^2 = AD \cdot BD$.

$$\because \angle CAD = \angle BAC, \therefore \text{Rt}\triangle CAD \sim \text{Rt}\triangle BAC$$
$$\therefore \frac{AC}{AD} = \frac{AB}{AC}, \text{ i.e. } AC^2 = AD \cdot AB.$$

The proof of $BC^2 = BD \cdot BA$ is similar.

Example 4. (Theorem on Centroid) For any triangle ABC, its three medians must intersect at one common point G, and each median is partitioned by G as two segments with ratio 2 : 1.

Solution Let G be the point of intersection of the medians AD and BE. From D introduce $DH \parallel BE$, such that DH intersects AC at H. By the midpoint theorem,

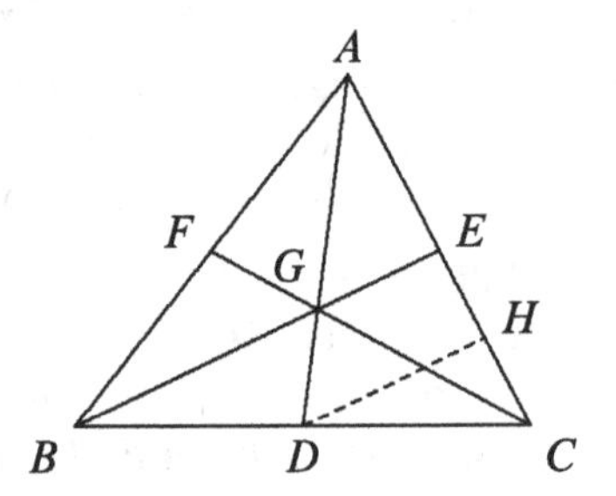

$$AE = EC = 2CH = 2HE.$$

$$\because \triangle AGE \sim \triangle ADH,$$

$$\therefore \frac{AG}{GD} = \frac{AE}{EH} = 2.$$

Similarly, $\dfrac{BG}{GE} = 2$.

Suppose that CF intersects AD at G', then similar to above proof, $\dfrac{AG'}{G'D} = 2$, hence $G = G'$. Thus AD, BE, CF are concurrent at G.

Example 5. (CHINA/1999) In $\triangle ABC$, AD is the median on BC, E is on AD such that $BE = AC$. The line BE intersects AC at F. Prove that $AF = EF$.

Solution From C introduce $CG \parallel AD$, intersecting the extension of BF at G.

$$\because \angle EAF = \angle FCG,$$
$$\angle AEF = \angle FGC,$$
$$\angle AFE = \angle GFC,$$
$$\therefore \triangle EAF \sim \triangle GCF \text{ (A.A.A.).}$$

$$\therefore \frac{AF}{EF} = \frac{FC}{FG} = \frac{AF + FC}{EF + FG} = \frac{AC}{EG}.$$

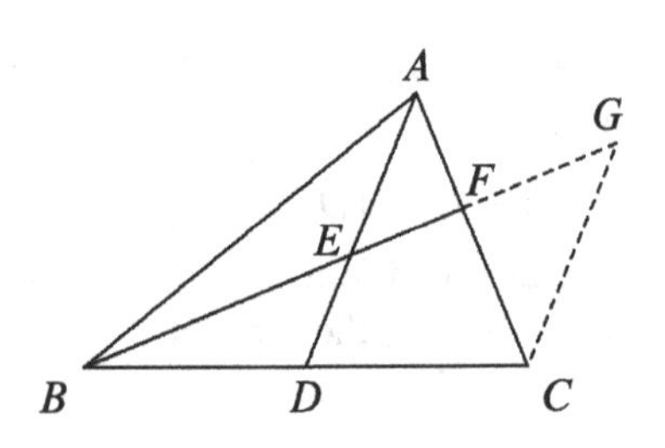

By the midpoint theorem, BE = EG,

$$\therefore EG = AC, AF = EF.$$

Example 6. (MOSCOW/1972) In $\triangle ABC$, AD, BE are medians on BC, AC respectively. If $\angle CAD = \angle CBE = 30°$, prove that $\triangle ABC$ is equilateral.

Solution $\because \triangle ADC \sim \triangle BEC$ (A.A.A.), therefore

$$\frac{AC}{BC} = \frac{DC}{EC} = \frac{2DC}{2EC} = \frac{BC}{AC},$$

$\therefore AC^2 = BC^2, \quad AC = BC.$

In $\triangle BEC$, $\angle EBC = 30°$, $EC = \frac{1}{2}BC$,

$\therefore \angle BEC = 90°, \angle C = 60°.$

$\therefore \triangle ABC$ is equilateral.

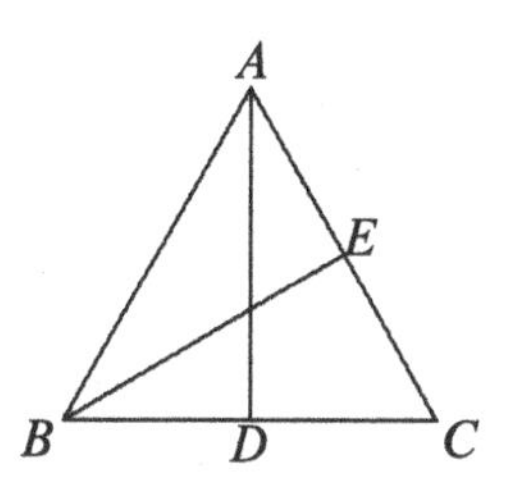

Example 7. In $\triangle ABC$, $\angle A : \angle B : \angle C = 1 : 2 : 4$. Prove that

$$\frac{1}{AB} + \frac{1}{AC} = \frac{1}{BC}.$$

Solution It suffices to show $\frac{AB + AC}{AB} = \frac{AC}{BC}$. To prove it we construct corresponding similar triangles as follows.

Extending AB to D such that $BD = AC$. Extending BC to E such that $AC = AE$. Connect DE, AE.

Let $\angle A = \alpha, \angle B = 2\alpha, \angle C = 4\alpha.$
Then $7\alpha = 180°.$
$\because \angle AEC = \angle ACE = 3\alpha,$
$\angle CAE = \alpha = \angle CAB,$
$\angle BAE = 2\alpha = \angle EBA.$
$\because \angle DBE = \angle BAE + \angle AEB = 5\alpha,$
$\therefore \angle EDA = \frac{1}{2}(180° - 5\alpha) = \alpha,$
$\therefore \triangle DAE \sim \triangle ABC$ (A.A.A.),

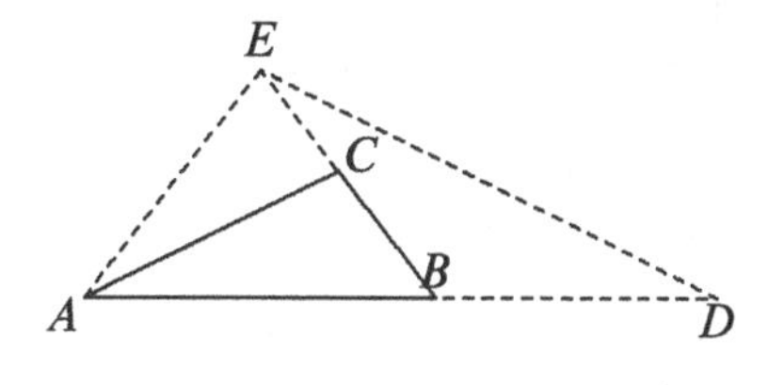

Thus, $\frac{AD}{AB} = \frac{AE}{BC}$, i.e. $\frac{AB + AC}{AB} = \frac{AC}{BC}$, as desired.

Example 8. (CHINA/1999) P is an inner point of $\triangle ABC$, $BC = a, CA = b, AB = c$. Through P introduce $IF \parallel BC, DG \parallel AB$ and $EH \parallel CA$ respectively, where I, H are on AB, D, E are on BC, F, G are on CA, as shown in the digram below. Given $DE = a', FG = b', HI = c'$, find the value of $\frac{a'}{a} + \frac{b'}{b} + \frac{c'}{c}$.

Solution From $IF \parallel BC, HE \parallel AC$ we have $\triangle HIP \sim \triangle ABC$, then

$$\frac{c'}{c} = \frac{IP}{a} = \frac{BD}{a}.$$

Since $\triangle GPF \sim \triangle ABC$,

$$\frac{b'}{b} = \frac{PF}{a} = \frac{EC}{a},$$

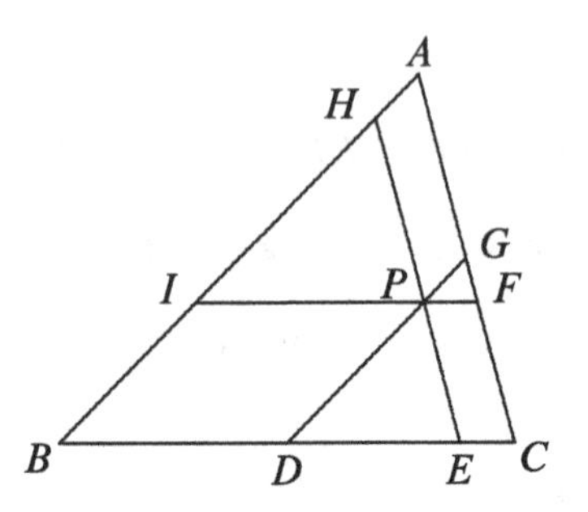

Therefore

$$\frac{a'}{a} + \frac{b'}{b} + \frac{c'}{c} = \frac{DE}{a} + \frac{BD}{a} + \frac{EC}{a}$$
$$= \frac{DE + BD + EC}{BC} = 1.$$

Example 9. In the given diagram, $\triangle PQR$ and $\triangle P'Q'R'$ are two congruent equilateral triangles. Denote the lengths of the sides of hexagon $ABCDEF$ by $AB = a_1,\ BC = b_1,\ CD = a_2,\ DE = b_2,\ EF = a_3,\ FA = b_3$. Prove that $a_1^2 + a_2^2 + a_3^2 = b_1^2 + b_2^2 + b_3^2$.

Solution For any two adjacent triangles outside the overlapping hexagon, say $\triangle Q'AB$ and $\triangle PAF$, we have

$$\angle P = \angle Q' = 60°,\quad \angle PAF = \angle Q'AB,$$

therefore $\triangle Q'AB \sim \triangle PAF$. Similarly, any two adjacent triangles are similar. Let

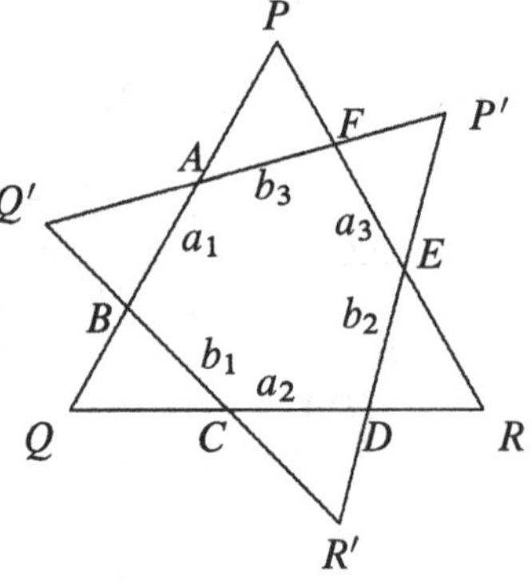

$$S_1 = [Q'AB],\quad S_2 = [QBC],\quad S_3 = [R'CD]$$
$$S_4 = [RDE],\quad S_5 = [P'FE],\quad S_6 = [PAF].$$

Then $S_1 + S_3 + S_5 = S_2 + S_4 + S_6$. Use A to denote the common sum above, then $\frac{b_1^2}{a_1^2} = \frac{S_2}{S_1},\ \frac{b_2^2}{a_1^2} = \frac{S_4}{S_1},\ \frac{b_3^2}{a_1^2} = \frac{S_6}{S_1}$. Adding them up, we obtain

$$\frac{b_1^2 + b_2^2 + b_3^2}{a_1^2} = \frac{A}{S_1},\quad \therefore a_1^2 \cdot A = (b_1^2 + b_2^2 + b_3^2)S_1.$$

Similarly, we have also

$$a_2^2 \cdot A = (b_1^2 + b_2^2 + b_3^2)S_2 \qquad \text{and} \qquad a_3^2 \cdot A = (b_1^2 + b_2^2 + b_3^2)S_3.$$

By adding these three equalities, we obtain

$$(a_1^2 + a_2^2 + a_3^2)A = (b_1^2 + b_2^2 + b_3^2)(S_1 + S_2 + S_3) = (b_1^2 + b_2^2 + b_3^2)A,$$

$$\therefore a_1^2 + a_2^2 + a_3^2 = b_1^2 + b_2^2 + b_3^2, \quad \text{as desired.}$$

Testing Questions (A)

1. (CHNMOL/1991) $ABCD$ is a trapezium with $AB \parallel CD$ and $AB < DC$. AC and BD intersect at E, $EF \parallel AB$, intersecting BC at F. Given that $AB = 20, CD = 80, BC = 100$, then EF is

   (A) 10, (B) 12, (C) 16, (D) 18.

2. (AHSME/1986) In $\triangle ABC$, $AB = 8, BC = 7, CA = 6$. Extend BC to P such that $\triangle PAB \sim \triangle PCA$, then the length of PC is

   (A) 7, (B) 8, (C) 9, (D) 10, (E) 11.

3. (CHINA/1990) In the isosceles right triangle ABC, $\angle B = 90°$, AD is the median on BC. Write $AB = BC = a$. If $BE \perp AD$, intersecting AC at E, and $EF \perp BC$ at F, then EF is

   (A) $\frac{1}{3}a$, (B) $\frac{1}{2}a$, (C) $\frac{2}{3}a$, (D) $\frac{2}{5}a$.

4. (CHINA/1997) ABC is an isosceles right triangle with $\angle C = 90°$, M, N are on AB such that $\angle MCN = 45°$. Write $AM = m, MN = x, BN = n$. Then the triangle formed by taking x, m, n as the lengths of it three sides is

   (A) an acute triangle; (B) a right triangle; (C) an obtuse triangle; (D) not determined.

5. In $\triangle ABC$, D is the midpoint of BC, E is on AC such that $AC = 3EC$. BE and AD intersect at G. Find $AG : GD$.

6. (CHINA/2000) Given that AD is the median on BC of $\triangle ABC$, E is a point on AD such that $AE = \frac{1}{3}AD$. The line CE intersects AB at F. If $AF =$ 1.2 cm, find the length of AB.

7. $ABCD$ is a rectangle with $AD = 2$, $AB = 4$. P is on AB such that $AP : PB = 2 : 1, CE \perp DP$ at E. Find CE.

8. Given that three congruent squares $ABEG, GEFH, HFCD$ are of side a. Prove that $\angle AFE + \angle ACE = 45°$.

9. (CHINA/1993) $\triangle ABC$ is equilateral, D is on BC such that $CD = 2BD$. If $CH \perp AD$ at H, prove that $\angle DBH = \angle DAB$.

10. In $\triangle ABC$, $\angle A = 2\angle B$. Prove that $AC^2 + AB \cdot AC = BC^2$.

Testing Questions (B)

1. (AIME/1984) A point P is chosen in the interior of $\triangle ABC$ such that when lines are drawn through P parallel to the sides of $\triangle ABC$, the resulting smaller triangles t_1, t_2, and t_3 in the figure, have areas 4, 9, and 49, respectively. Find the area of $\triangle ABC$.

2. (APMO/1993) Let $ABCD$ be a quadrilateral such that all sides have equal length and angle ABC is 60°. Let l be a line passing through D and not intersecting the quadrilateral (except at D). Let E and F be the points of intersection of l with AB and BC respectively. Let M be the point of intersection of CE and AF. Prove that $CA^2 = CM \cdot CE$.

3. (CHINA/1997) In the $\triangle ABC$, D, E are on BC, AC respectively, such that $\frac{BD}{DC} = \frac{2}{3}, \frac{AE}{EC} = \frac{3}{4}$. Find the value of $\frac{AF}{FD} \cdot \frac{BF}{FE}$.

4. (CHINA/1979) In a Rt$\triangle ABC$, $\angle C = 90°$, BE is the angle bisector of $\angle B$, $CD \perp AB$ at D and CD intersects BE at O. Through O introduce $FG \parallel AB$ such that FG intersects AC, BC at F, G respectively. Prove that $AF = CE$.

5. (CHINA/1998) In the quadrilateral $ABCD$, AC and BD intersect at O, the line l is parallel to BD, intersecting the extensions of AB, DC, BC, AD and AC at the points M, N, R, S and P respectively. Prove that $PM \cdot PN = PR \cdot PS$.

Lecture 14

Areas of Triangles and Applications of Area

Basic formulae for area of a triangle

Note: The area of a triangle UVW is denoted by $[UVW]$ hereafter.

Theorem I. *For* $\triangle ABC$, $[ABC] = \frac{1}{2}a \cdot h_a = \frac{1}{2}b \cdot h_b = \frac{1}{2}c \cdot h_c$, *where* $BC = a, CA = b, AB = c$ *and* h_a, h_b, h_c *are altitudes on* BC, CA, AB *respectively.*

Theorem II. *(Heron's Formula) For* $\triangle ABC$, $[ABC] = \sqrt{s(s-a)(s-b)(s-c)}$, *where* $s = \frac{1}{2}(a+b+c)$.

Proof. For $\triangle ABC$, let $BC = a, CA = b, AB = c$, and $AD = h$, where $AD \perp BC$ at D. Let $CD = x$, then

$$c^2 - (a-x)^2 = h^2 = b^2 - x^2,$$
$$c^2 - a^2 + 2ax = b^2,$$
$$\therefore x = \frac{a^2 + b^2 - c^2}{2a}.$$

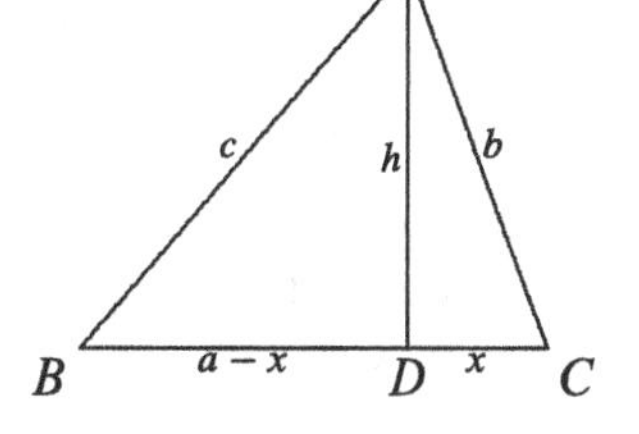

Therefore

$$\begin{aligned} h^2 &= b^2 - \left(\frac{a^2+b^2-c^2}{2a}\right)^2 = \frac{1}{4a^2}\left[(2ab)^2 - (a^2+b^2-c^2)^2\right] \\ &= \frac{(2ab+a^2+b^2-c^2)(2ab-a^2-b^2+c^2)}{4a^2} \\ &= \frac{[(a+b)^2-c^2][c^2-(a-b)^2]}{4a^2} \end{aligned}$$

$$= \frac{(a+b+c)(a+b-c)(c+a-b)(c-a+b)}{4a^2}$$
$$= \frac{16}{4a^2}s(s-c)(s-b)(s-a) = \frac{4s(s-a)(s-b)(s-c)}{a^2}.$$

$\therefore [ABC]^2 = \left(\frac{ha}{2}\right)^2 = s(s-a)(s-b)(s-c).$

Thus, the conclusion is proven. □

Note: If $\angle C$ is obtuse, then $c^2 > a^2 + b^2$, so $x < 0$, but the proof remains unchanged.

Comparison of areas of triangles

(I) For each triangle, let $S = h \cdot b$, where b is one side and h is the height on the side. then the ratio of areas of any two triangles is equal to the ratio of corresponding two Ss.

(II) For two triangles with equal bases, the ratio of their areas is equal to the ratio of their heights on the bases. Hence the area of a triangle does not change when a vertex of it moves on a line parallel to its opposite side.

(III) For two triangles with equal height, the ratio of their areas is equal to the ratio of their bases.

(IV) If two triangles have a pair of equal angles, then the ratio of their areas is equal to the ratio of the products of the two sides of the equal angles.

(V) If two triangles have a pair of supplementary angles, then the ratio of their areas is equal to the ratio of the products of the two sides of the supplementary angles.

There are two kinds of questions to be discussed in this chapter. One is those for finding areas or discussing questions involving areas. The other is those able to be solved by considering areas. Below some examples of these two kinds of questions are given.

Examples

Example 1. (SMO/1988) Suppose area of $\triangle ABC = 10$ cm^2, $AD = 2$ cm, $DB = 3$ cm and area of $\triangle ABE$ is equal to area of quadrilateral $DBEF$. Then area of $\triangle ABE$ equals

(A) 4 cm^2 (B) 5 cm^2 (C) 6 cm^2 (D) 7 cm^2 (E) 8 cm^2.

Solution Connect DE. Since $[ABE] = [DBEF]$, we have

$$[ADE] = [ABE] - [DBE]$$
$$= [DBEF] - [DBE] = [FDE],$$
$$\therefore AC \parallel DE,$$
$$\therefore CE : EB = AD : DB = 2 : 3.$$
$$\because \frac{[ABE]}{[ABC]} = \frac{BE}{BC} = \frac{3}{5},$$
$$\therefore [ABE] = \frac{3}{5} \cdot [ABC] = 6.$$

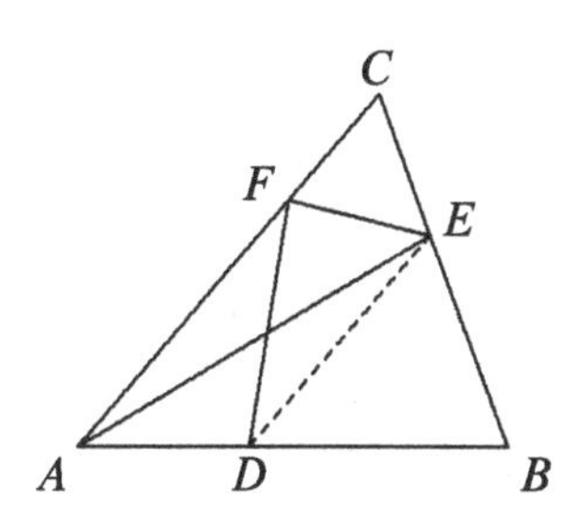

Thus, the answer is (C).

Example 2. In the given diagram, $ABCD$ is a convex quadrilateral with $[ABC] > [DAC]$. Find a point M on the segment BC such that AM partitions $ABCD$ to two parts with equal areas.

Solution We need to change the shape of the graph from quadrilateral to a triangle, keeping its area unchanged. From D introduce $ED \parallel AC$ such that DE intersects the extension of line BC at E. Then

$$[DAC] = [EAC], \quad \therefore [ABCD] = [EAB].$$

Now taking M be the midpoint of BE, then

$$[ABM] = [AEM] = \tfrac{1}{2}[ABCD]$$
$$= [AMCD].$$

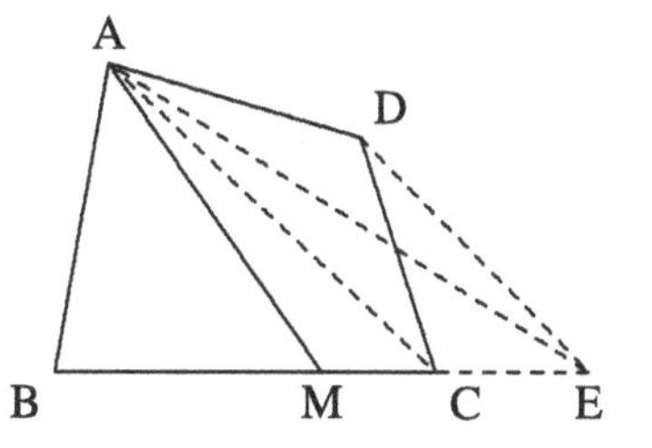

Example 3. In $\triangle ABC$, D, E are on BC and CA respectively, and $BD : DC = 3 : 2, AE : EC = 3 : 4$. AD and BE intersect at M. Given that the area of $\triangle ABC$ is 1, find area of $\triangle BMD$.

Solution From E introduce $EN \parallel AD$, intersecting BC at N. Since

$$\frac{DN}{NC} = \frac{AE}{EC} = \frac{3}{4}, \frac{BD}{DC} = \frac{3}{2},$$
$$[ABE] = \frac{3}{7}[ABC] = \frac{3}{7},$$
$$\therefore [BEC] = \frac{4}{7}[ABC] = \frac{4}{7}.$$
$$\because BD : DN : NC = 21 : 6 : 8,$$
$$\therefore BN : NC = 27 : 8 \text{ and}$$

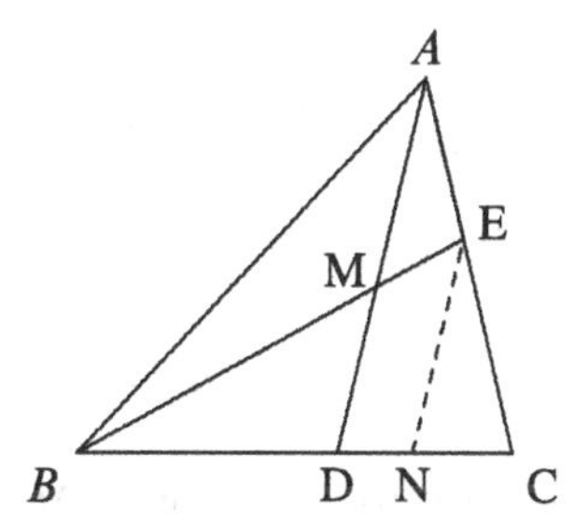

$$BD : BN = 21 : 27 = 7 : 9, \ \ [BEN] = \frac{27}{35}[BEC] = \frac{27}{35} \cdot \frac{4}{7},$$
$$[BMD] = \left(\frac{7}{9}\right)^2 [BEN] = \frac{7^2 \cdot 27 \cdot 4}{9^2 \cdot 35 \cdot 7} = \frac{4}{15}.$$

Example 4. (USAMO/1972) A convex pentagon $ABCDE$ has the following property: the five triangles ABC, BCD, CDE, DEA, EAB have same area 1. Prove that all such pentagons have an equal area.

Solution As shown in the right diagram, the equality $[EAB] = [CAB]$ yields $EC \parallel AB$. Similarly, we have

$$AD \parallel BC, \ \ BE \parallel CD, \ \ AC \parallel DE, \ \ BD \parallel AE.$$

Let $[BPC] = x$. Then $[DPC] = 1 - x$ and

$$\frac{[BPC]}{[DPC]} = \frac{BP}{PD} = \frac{[EBP]}{[EPD]},$$

so it follows that $\frac{x}{1-x} = \frac{1}{x}$,

$$\therefore x^2 + x - 1 = 0, \ \ x = \frac{\sqrt{5}-1}{2},$$
$$\therefore [ABCDE] = 3 + x = \frac{5+\sqrt{5}}{2}.$$

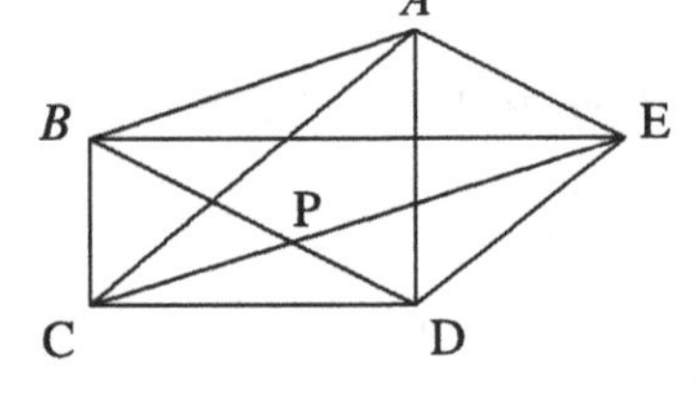

Example 5. In a quadrilateral $ABCD$, the points E, F are on AB and H, G are on DC such that $AE = EF = FB$ and $DH = HG = GC$. Prove that $[EFGH] = \frac{1}{3}[ABCD]$.

Solution Connect HF, AH, AC, FC.

$$\because [HEF] = [HEA] \ \text{and} \ [FGH] = [FGC],$$
$$\therefore [EFGH] = \frac{1}{2}[HAFC].$$

It suffices to show that $[ADH] + [CFB] = \frac{1}{3}[ABCD]$.

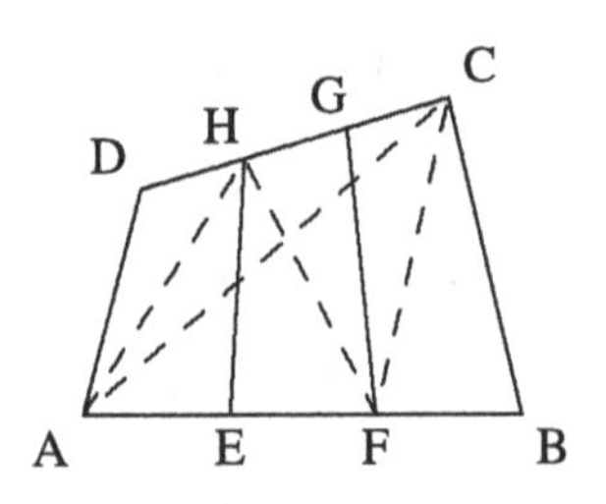

From $DH = \frac{1}{3}CD$ and $FB = \frac{1}{3}AB$, we have

$$[ADH] + [CFB] = \frac{1}{3}([DAC] + [BAC]) = \frac{1}{3}[ABCD],$$

$$\therefore [EFGH] = \frac{1}{2}[HAFC] = \frac{1}{2} \cdot \frac{2}{3}[ABCD] = \frac{1}{3}[ABCD].$$

Example 6. (AIME/1985) As shown in the figure, triangle ABC is divided into six smaller triangles by lines drawn from the vertices through a common interior point. The areas of four of these triangles are as indicated. Find the area of triangle ABC.

Solution $\dfrac{[CAP]}{[FAP]} = \dfrac{CP}{FP} = \dfrac{[CBP]}{[FBP]}$ yields

$$\frac{84+y}{40} = \frac{x+35}{30}, \qquad (14.1)$$

and $\dfrac{[CAP]}{[CDP]} = \dfrac{AP}{DP} = \dfrac{[BAP]}{[BDP]}$ yields

$$\frac{84+y}{x} = \frac{70}{35} = 2. \qquad (14.2)$$

By $\dfrac{(1)}{(2)}$, it follows that $\dfrac{x}{40} = \dfrac{x+35}{60}$,

$\therefore 3x = 2x + 70$, i.e. $x = 70$. Then by (2), $y = 140 - 84 = 56$.

Thus, $[ABC] = 84 + 56 + 40 + 30 + 35 + 70 = 315$.

Example 7. If from any fixed inner point P of $\triangle ABC$ introduce $PD \perp BC$ at D, $PE \perp CA$ at E and $PF \perp AB$ at F. Prove that $\dfrac{PD}{h_a} + \dfrac{PE}{h_b} + \dfrac{PF}{h_c} = 1$, where h_a, h_b, h_c are the heights of $\triangle ABC$ introduced from A, B, C to their opposite sides, respectively.

Solution From

$$\frac{PD}{h_a} = \frac{[PBC]}{[ABC]},$$

$$\frac{PE}{h_b} = \frac{[PCA]}{[ABC]},$$

$$\frac{PF}{h_c} = \frac{[PAB]}{[ABC]},$$

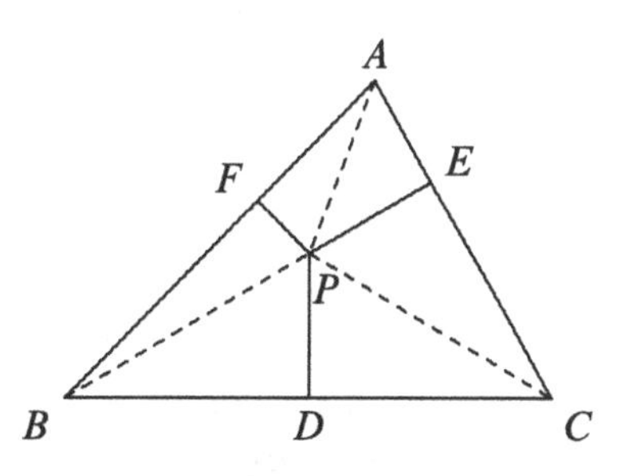

it follows that

$$\frac{PD}{h_a} + \frac{PE}{h_b} + \frac{PF}{h_c} = \frac{[PBC] + [PCA] + [PAB]}{[ABC]} = 1.$$

Example 8. (AUSTRALIA/1991) In $\triangle ABC$, M is the midpoint of BC, P, R are on AB, AC respectively, Q is the point of intersection of AM and PR. If Q is the midpoint of PR, prove that $PR \parallel BC$.

Solution From that Q, M are the midpoints of PR and BC respectively,

$$[APQ] = [ARQ], \quad [ABM] = [ACM],$$

$$\therefore \frac{[APQ]}{[ABM]} = \frac{[ARQ]}{[ACM]},$$

$$\therefore \frac{AP \cdot AQ}{AB \cdot AM} = \frac{AQ \cdot AR}{AM \cdot AC},$$

i.e. $\dfrac{AP}{AB} = \dfrac{AR}{AC}$, $\therefore PQ \parallel BC$.

Example 9. (CHINA/1992) In the given diagram below, $ABCD$ is a parallelogram, E, F are two points on the sides AD and DC respectively, such that $AF = CE$. AF and CE intersect at P. Prove that PB bisects $\angle APC$.

Solution Connect BE, BF, make $BU \perp AF$ at U and $BV \perp CE$ at V. Then

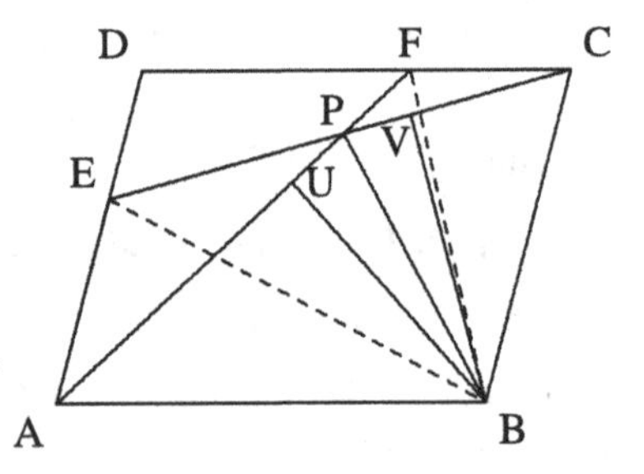

$$[BAF] = [BCE] = \frac{1}{2}[ABCD].$$

Further, since $AF = CE$, we have

$$BU = BV, \ \therefore \triangle BPU \cong \triangle BPV,$$

$$\therefore \angle BPA = \angle BPU = \angle BPV = \angle BPC.$$

Testing Questions (A)

1. (CHINA/1993) When extending the sides AB, BC, CA of $\triangle ABC$ to B', C', A' respectively, such that $AB' = 2AB, CC' = 2BC, AA' = 3CA$. If area of $\triangle ABC$ is 1, find the area of $\triangle A'B'C'$.

2. (CHINA/1998) $ABCD$ is a rectangle. $AD = 12, AB = 5$. P is a point on AD, $PE \perp BD$ at E, $PF \perp AC$ at F. Find $PE + PF$.

3. (CHINA/1996) Given that the point P is outside the equilateral triangle ABC

but inside the region of $\angle ABC$. If the distances from P to BC, CA, AB are h_1, h_2 and h_3 respectively, and $h_1 - h_2 + h_3 = 6$, find the area of $\triangle ABC$.

4. (CHINA/1996) Let the heights on three sides of $\triangle ABC$ be h_a, h_b, h_c respectively, and $2b = a + c$. Prove that $\frac{2}{h_b} = \frac{1}{h_a} + \frac{1}{h_c}$.

5. (CHINA/2000) In $\triangle ABC$, D, E, F are on the sides BC, CA, AB respectively, such that they are concurrent at a point G, $BD = 2CD$, the areas $S_1 = [GEC] = 3$, $S_2 = [GCD] = 4$. Find the area of $\triangle ABC$.

6. (CHINA/1958) Let AD, BE, CF be the three angle bisectors of the triangle ABC, prove that the ratio of area of $\triangle DEF$ to area of $\triangle ABC$ is equal to $\frac{2abc}{(a+b)(b+c)(c+a)}$, where $a = BC, b = CA$ and $c = AB$.

7. In a trapezium $ABCD$, $AD \parallel BC$, the extensions of BA and CD intersect at E. Make $EF \parallel BD$ where EF intersects the extension of CB at F. On the extension of BC take G such that $CG = BF$. Prove $EG \parallel AC$.

8. (AIME/1988) Let P be an interior point of triangle ABC and extend lines from the vertices through P to the opposite sides. Let a, b, c, and d denote the lengths of the segments indicated in the figure below. Find the product abc if $a + b + c = 43$ and $d = 3$.

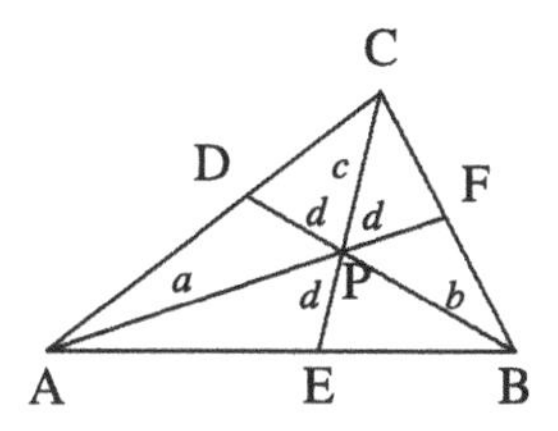

9. (CHNMOL/1998) In the isosceles right triangle ABC, $AB = 1, \angle A = 90°$, E is the midpoint of the leg AC. The point F is on the base BC such that $EF \perp BE$. Find the area of $\triangle CEF$.

Testing Questions (B)

1. (IMO/Shortlist/1989) In the convex quadrilateral $ABCD$, the midpoints of BC and AD are E and F respectively. Prove that $[EDA] + [FBC] = [ABCD]$.

2. (JAPAN/1991) Given that G is the centroid of $\triangle ABC$, $GA = 2\sqrt{3}, GB = 2\sqrt{2}, GC = 2$. Find the area of $\triangle ABC$.

3. (Ceva's Theorem) P is an inner point of $\triangle ABC$. Extend the lines AP, BP, CP to intersect the opposite side at D, E, F respectively, Then

$$\frac{BD}{DC} \cdot \frac{CE}{EA} \cdot \frac{AF}{FB} = 1.$$

4. (AIME/1992) In triangle ABC, A', B', and C' are on the sides BC, CA, and AB, respectively. Given that AA', BB', and CC' are concurrent at the point O, and that $\frac{AO}{OA'} + \frac{BO}{OB'} + \frac{CO}{OC'} = 92$, find $\frac{AO}{OA'} \cdot \frac{BO}{OB'} \cdot \frac{CO}{OC'}$.

5. (AIME/1989) Point P is inside $\triangle ABC$. Line segments APD, BPE, and CPF are drawn with D on BC, E on CA, and F on AB (see the figure below). Given that $AP = 6$, $BP = 9$, $PD = 6$, $PE = 3$, and $CF = 20$, find the area of $\triangle ABC$.

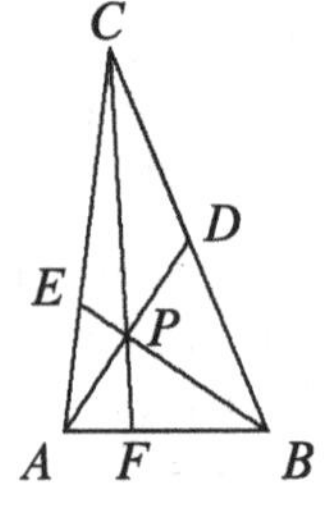

Lecture 15

Divisions of Polynomials

(I) **Long Division**: When a whole number n is divided by a non-zero whole number m, a quotient q and a remainder r can be obtained. The result can be written in the form $n = mq + r$, where $0 \le r < m$, and the process for getting the expression is called **division with remainder**.
In the division of polynomials, when a polynomial $f(x)$ is divided by a non-zero polynomial $g(x)$ under the usual division (**long division**), there is a quotient $q(x)$ and a remainder $r(x)$, where the degree of f is sum of the degrees of g and q when q is not zero polynomial (or $f = r$ if $q \equiv 0$), and the degree of r is less than that of g. The result can be written in the form

$$f(x) = g(x) \cdot q(x) + r(x).$$

(II) **Synthetic Division**. When the divisor $g(x) = x - a$, where a is a constant, then $q(x)$ is a polynomial with degree $n - 1$ if the degree of $f(x)$ is n, and the remainder r is a constant. Write

$$f(x) = a_n x^n + a_{n-1} x^{n-1} + \cdots + a_1 x + a_0,$$
$$q(x) = b_{n-1} x^{n-1} + b_{n-2} x^{n-1} + \cdots + b_1 x + b_0.$$

Since

$$\begin{aligned} f(x) &= (x-a)(b_{n-1}x^{n-1} + b_{n-2}x^{n-1} + \cdots + b_1 x + b_0) + r \\ &= b_{n-1}x^n + (b_{n-2} - ab_{n-1})x^{n-1} + (b_{n-3} - ab_{n-2})x^{n-2} \\ &\quad + \cdots + (b_0 - ab_1)x - ab_0 + r, \end{aligned}$$

by the comparison of the coefficients of $f(x)$, we have

$$b_{n-1} = a_n, \ b_{n-2} = a_{n-1} + ab_{n-1}, \ \cdots, b_0 = a_1 + ab_1, \ r = a_0 + ab_0.$$

Thus, the following operation table is obtained:

$$
\begin{array}{c|cccccc|c}
a & a_n & a_{n-1} & a_{n-2} & \cdots & a_1 & a_0 \\
 & +) & ab_{n-1} & ab_{n-2} & \cdots & ab_1 & ab_0 \\
\hline
 & b_{n-1} & b_{n-2} & b_{n-3} & \cdots & b_0 & \multicolumn{1}{|c}{r}
\end{array}
$$

(III) **Remainder Theorem and Factor Theorem**

Theorem I. *The Remainder Theorem: For any constant a, a polynomial $f(x)$ of degree $n \geq 1$ can be written in the form $f(x) = (x-a)q(x) + r$, $q(x)$ is a polynomial of degree $n-1$, and $r = f(a)$.*
For any given constants a and b, a polynomial $f(x)$ of degree $n \geq 2$ can be written in the form $f(x) = (x-a)(x-b)q(x) + r(x)$, where $q(x)$ is a polynomial of degree $n-2$, and the remainder $r(x)$ is zero polynomial or has degree 1 or zero.

Theorem II. *The Factor Theorem: A polynomial $f(x)$ has the factor $(x-a)$ if and only if $f(a) = 0$.*

Theorem III. *For a polynomial with integral coefficients $f(x) = a_n x^n + a_{n-1}x^{n-1} + \cdots + a_1 x + a_0$, if it has the factor $x - \frac{q}{p}$, where p, q are two relatively prime integers, then q is a factor of a_0, and p is a factor of a_n.*

Proof. By the factor theorem, $f\left(\frac{q}{p}\right) = 0$, therefore

$$
\begin{aligned}
0 &= a_n\left(\frac{q}{p}\right)^n + a_{n-1}\left(\frac{q}{p}\right)^{n-1} + \cdots + a_1\left(\frac{q}{p}\right) + a_0 \\
0 &= a_n q^n + a_{n-1}pq^{n-1} + \cdots + a_1 p^{n-1}q + a_0 p^n, \\
-a_0 p^n &= (a_n q^{n-1} + a_{n-1}pq^{n-2} + \cdots + a_1 p^{n-1})q,
\end{aligned}
$$

so $q \mid a_0 p^n$, which implies that $q \mid a_0$ since $(p,q) = 1$. Similarly, we have

$$
-a_n q^n = (a_{n-1}q^{n-1} + a_{n-1}pq^{n-2} + \cdots + a_1 p^{n-2}q + a_{)}p^{n-1})p,
$$

so $p \mid a_n q^n$, which implies $p \mid a_n$ since $(p,q) = 1$. □

(IV) **The Factorization of Symmetric or Cyclic Polynomials**. A polynomial of multi-variables is called **symmetric** if the polynomial does not change when taking any permutation transformation to its variables. For example, $x + y + z, x^2 + y^2 + z^2, xyz$, etc..

A symmetric expression containing two variables x and y can be always factorized as a product of factors expressed in terms of the basic symmetric expressions $(x+y), xy$; a symmetric expression containing three variables x, y, z can be always factorized as a product of factors expressed in terms of the basic symmetric expressions $(x+y+z), xy+yz+zx, xyz$.
A polynomial of multi-variables is called **cyclic** if after changing all its variables cyclically the resulting polynomial does not change. For example, $xy+yz+zx, x^2y+y^2z+z^2x, (x+y)(y+z)(z+x)$, etc.. A symmetric polynomial must be cyclic, but the inverse is not always true.
If a cyclic polynomial has a factor, then the expressions obtained by cyclically changing its variables of the factor are all factors of the polynomial, since the factorization of the polynomial is also cyclic. Based on this feature, we can consider one of the variables as the principal one, and the others as constants at the moment, so that we get a polynomial of single variable. Then it is easy to find a factor by using the factor theorem, and using above cyclic feature of the factors, we can obtain the other factors at once. Finally, if there are a few constant coefficients to be determined, then the coefficient-determining method (cf. Lecture VI) is useful for this.

Examples

Example 1.
(i) When $f(x)=3x^2+5x-7$ is divided by $x+2$, find the quotient and remainder by long division.
(ii) When $f(x)=3x^4-5x^3+x^2+2$ is divided by x^2+3, find the quotient and remainder by long division.

Solution By the following long division:

$$\begin{array}{r|l} & 3x-1 \\ \hline x+2 & 3x^2+5x-7 \\ & 3x^2+6x \\ \hline & \quad -x-7 \\ & \quad -x-2 \\ \hline & \qquad -5 \end{array}$$

$$\begin{array}{r|l} & 3x^2-5x-8 \\ \hline x^2+3 & 3x^4-5x^3+x^2+2 \\ & 3x^4+9x^2 \\ \hline & \quad -5x^3-8x^2+2 \\ & \quad -5x^3-15x \\ \hline & \quad -8x^2+15x+2 \\ & \quad -8x^2 \qquad -24 \\ \hline & \qquad 15x+26 \end{array}$$

Therefore
(i) $q(x)=3x-1$, $r=-5$. (ii) $q(x)=3x^2-5x-8$, $r(x)=15x+26$.

Note: Please compare the long divisions for whole numbers and that for polynomials, what are the same features and what are the distinctions?

Example 2. Use synthetic division to find the quotient and remainder of the polynomial $2x^4 - 3x^3 - x^2 + 5x + 6$ when it is divided by $x + 1$.

Solution By using synthetic division, the following result is obtained:

$$\begin{array}{r|rrrr|r}
-1 & 2 & -3 & -1 & 5 & 6 \\
 & & -2 & 5 & -4 & -1 \\
\hline
 & 2 & -5 & 4 & 1 & 5 \\
\cline{6-6}
\end{array}$$

Thus, $q(x) = 2x^3 - 5x^2 + 4x + 1, \quad r = 5.$

When a polynomial $f(x)$ is divided by $g(x) = ax + b$, where $a \neq 1$ and $a \neq 0$, the synthetic division still works, since

$$f(x) = (ax + b)q(x) + r = (x + \frac{b}{a}) \cdot (aq(x)) + r,$$

so we let the divisor be $x + \dfrac{b}{a}$ to use the synthetic division first, after getting $ag(x)$ and r, the $q(x)$ and r are obtained at once.

Example 3. Find the quotient and remainder of the polynomial $6x^4 - 7x^3 - x^2 + 8$ when it is divided by $2x + 1$.

Solution By using synthetic division to carry out the division $(6x^4 - 7x^3 - x^2 + 8) \div (x + \frac{1}{2})$:

$$\begin{array}{r|rrrr|r}
-\frac{1}{2} & 6 & -7 & -1 & 0 & 8 \\
 & & -3 & 5 & -2 & 1 \\
\hline
 & 6 & -10 & 4 & -2 & 9 \\
\cline{6-6}
\end{array}$$

therefore $2q(x) = 6x^3 - 10x^2 + 4x - 2, r = 9$, so $q(x) = 3x^3 - 5x^2 + 2x - 1, r = 9$. Note that the remainder r is not effected by the change of divisor.

Example 4. If a polynomial $f(x)$ has remainders 3 and 5 when divided by $x - 1$ and $x - 2$ respectively, find the remainder when $f(x)$ is divided by $(x - 1)(x - 2)$.

Solution From the Remainder Theorem,

$$f(x) = (x-1)q_1(x) + 3 \text{ and } q_1(x) = (x-2)q_2(x) + r,$$

where $q_1(x)$ is the quotient of f when divided by $x-1$, q_2, r are the quotient and remainder of q_1 respectively when divided by $x-2$. Then

$$f(x) = (x-1)[(x-2)q_2(x) + r] + 3 = (x-1)(x-2)q_2(x) + r(x-1) + 3.$$

By the Remainder Theorem, we have $5 = f(2) = r(2-1) + 3 = r + 3$, so $r = 2$. Thus, the remainder of f when divided by $(x-1)(x-2)$ is $2x+1$.

Example 5. If a polynomial $f(x)$ is divisible by both $x-a$ and $x-b$, where a, b are two different constants, prove that $f(x)$ must be divisible by $(x-a)(x-b)$.

Solution Similar to the preceding question, let

$$f(x) = (x-a)q_1(x) \quad \text{and} \quad q_1(x) = (x-b)q_2(x) + r,$$

then

$$f(x) = (x-a)[(x-b)q_2(x) + r] = (x-a)(x-b)q_2(x) + r(x-a).$$

By the Factor Theorem,

$$0 = f(b) = r(b-a), \quad \therefore r = 0.$$

Thus, $f(x) = (x-a)(x-b)q_2(x)$, the conclusion is proven.

Example 6. Factorize $f(x) = x^4 + x^3 - 7x^2 - x + 6$.

Solution From Theorem IV, if $f(x) = 0$ has rational roots, then they must be integral roots, and are in the set $S = \{\pm 1, \pm 2, \pm 3, \pm 6\}$. Since $f(1) = f(-1) = 0$, by the factor theorem, f has factors $(x-1)$ and $(x+1)$. To check the other numbers in S, by synthetic division,

$$\begin{array}{r|rrrrr} 2 & 1 & 1 & -7 & -1 & 6 \\ & & 2 & 6 & -2 & -6 \\ \hline & 1 & 3 & -1 & -3 & \boxed{0} \end{array}$$

therefore the quotient $q_1(x)$ of $f(x)$ when divided by $(x-2)$ is $x^3 + 3x^2 - x - 3$, remainder is 0, so $x-2$ is the third factor of $f(x)$. Next, we check the factor

$x + 3$. By synthetic division,

$$\begin{array}{r|rrrr} -3 & 1 & 3 & -1 & -3 \\ & & -3 & 0 & 3 \\ \hline & 1 & 0 & -1 & \boxed{0} \end{array}$$

Thus, the quotient $q_2(x)$ of $x^3 + 3x^2 - x - 3$ when divided by $x + 3$ is $x^2 - 1$, and remainder is 0. So $x + 3$ is the fourth factor. Thus,

$$f(x) = (x - 1)(x + 1)(x - 2)(x + 3).$$

Example 7. Factorize the symmetric expression $(x + y + z)^5 - x^5 - y^5 - z^5$.

Solution By taking x as the principal variable, we define $f(x) = (x + y + z)^5 - x^5 - y^5 - z^5$. Since

$$f(-y) = z^5 - (-y)^5 - y^5 - z^5 = 0,$$

$(x + y)$ is a factor of the original expressions, and so are the expressions $(y + z)$ and $(z + x)$. Assume that

$$(x+y+z)^5-x^5-y^5-z^5 = (x+y)(y+z)(z+x)[A(x^2+y^2+z^2)+B(xy+yz+zx)],$$

then the comparison of the coefficients of x^4y on both sides leads to $A = 5$. Let $x = y = z = 1$, it follows that $243 - 3 = 8[15 + 3B]$, so $B = 5$. Thus,

$$(x+y+z)^5-x^5-y^5-z^5 = 5(x+y)(y+z)(z+x)(x^2+y^2+z^2+xy+yz+zx).$$

Example 8. (MOSCOW/1940) Factorize $(b - c)^3 + (c - a)^3 + (a - b)^3$.

Solution Taking a as the principal variable and let $f(a) = (b - c)^3 + (c - a)^3 + (a - b)^3$. Then

$$f(b) = (b - c)^3 + (c - b)^3 = 0,$$

so $(a - b)$ is a factor of the original expression. Hence $(b - c)$ and $(c - a)$ are also the factors. Thus

$$(b - c)^3 + (c - a)^3 + (a - b)^3 = A(b - c)(c - a)(a - b),$$

where A is a constant to be determined. Let $a = 2, b = 1, c = 0$ on both sides, then $-6 = -2A$, so $A = 3$. Thus,

$$(b - c)^3 + (c - a)^3 + (a - b)^3 = 3(b - c)(c - a)(a - c).$$

Note: Considering $(b-c)+(c-a)+(a-b)=0$, the above result can be obtained at once by using the formula

$$u^3+v^3+w^3-3uvw=(u+v+w)(u^2+v^2+w^2-uv-vw-wu).$$

Example 9. Factorize $a^3(b-c)+b^3(c-a)+c^3(a-b)$.

Solution Taking a as the principal variable and $f(a)=a^3(b-c)+b^3(c-a)+c^3(a-b)$, we have

$$f(b)=b^3(b-c)+b^3(c-b)=0,$$

so $(a-b),(b-c),(c-a)$ are all factors of the original expression. Then

$$a^3(b-c)+b^3(c-a)+c^3(a-b)=A(a+b+c)(a-b)(b-c)(c-a).$$

Let $a=2, b=1, c=0$, then $8-2=-6A$, i.e. $A=-1$. Thus

$$a^3(b-c)+b^3(c-a)+c^3(a-b)=(a+b+c)(b-c)(a-c)(a-b).$$

Testing Questions (A)

1. Find, by long division and synthetic division respectively, the quotient and remainder of $3x^3-5x+6$ when it is divided by $x-2$.
2. Use synthetic division to carry out the division $(-6x^4-7x^2+8x+9)\div(2x-1)$.
3. Given that $f(x)=x^4+3x^3+8x^2-kx+11$ is divisible by $x+3$, find the value of k.
4. Given that $f(x)=x^4-ax^2-bx+2$ is divisible by $(x+1)(x+2)$, find the values of a and b.
5. Given that a polynomial $f(x)$ has remainders $1,2,3$ when divided by $(x-1),(x-2),(x-3)$, respectively. Find the remainder of $f(x)$ when it is divided by $(x-1)(x-2)(x-3)$.
6. If $x^5-5qx+4r$ is divisible by $(x-2)^2$, find the values of q and r.
7. Given that $f(x)$ is a polynomial of degree 3, and its remainders are $2x-5$ and $-3x+4$ when divided by x^2-1 and x^2-4 respectively. Find the $f(x)$.
8. Factorize $x^3+7x^2+14x+8$.

9. Factorize $x^4 + y^4 + (x + y)^4$.

10. Factorize $xy(x^2 - y^2) + yz(y^2 - z^2) + zx(z^2 - x^2)$.

Testing Questions (B)

1. Given that $f(x) = x^2 + ax + b$ is a polynomial with integral coefficients. If f is a common factor of polynomials $g(x) = x^4 - 3x^3 + 2x^2 - 3x + 1$ and $h(x) = 3x^4 - 9x^3 + 2x^2 + 3x - 1$, find $f(x)$.

2. For any non-negative integers m, n, p, prove that the polynomial $x^{3m} + x^{3n+1} + x^{3p+2}$ has the factor $x^2 + x + 1$.

3. Given that $f(x)$ is a polynomial with real coefficients. If there are distinct real numbers a, b, c, such that the remainders of $f(x)$ are a, b, c when f is divided by $(x - a), (x - b), (x - c)$ respectively, prove that $f(x) - x$ has the factor $(x - a)(x - b)(x - c)$.

4. Factor $(y^2 - z^2)(1 + xy)(1 + xz) + (z^2 - x^2)(1 + yz)(1 + yx) + (x^2 - y^2)(1 + zx)(1 + zy)$.

5. When $f(x) = x^3 + 2x^2 + 3x + 2$ is divided by $g(x)$ which is a polynomial with integer coefficients, the quotient and remainder are both $h(x)$. Given that h is not a constant, find g and h.

Solutions to Testing Questions

Solutions To Testing Questions

Solutions to Testing Questions 1

Testing Questions (1-A)

1. -1.

2. 0.

3. 1.

$$2009\left(1-\frac{1}{2}\right)\left(1-\frac{1}{3}\right)\cdots\left(1-\frac{1}{2009}\right)$$
$$=2009\cdot\frac{1}{2}\cdots\frac{2}{3}\cdot\frac{3}{4}\cdots\cdots\frac{2008}{2009}=1.$$

4.
$$\frac{1}{5\times7}+\frac{1}{7\times9}+\frac{1}{9\times11}+\frac{1}{11\times13}+\frac{1}{13\times15}$$
$$=\frac{1}{2}\left[\left(\frac{1}{5}-\frac{1}{7}\right)+\left(\frac{1}{7}-\frac{1}{9}\right)+\cdots+\left(\frac{1}{13}-\frac{1}{15}\right)\right]$$
$$=\frac{1}{2}\left(\frac{1}{5}-\frac{1}{15}\right)=\frac{1}{15}.$$

5. $\frac{1}{10}+\frac{1}{40}+\frac{1}{88}+\frac{1}{154}+\frac{1}{238}=\frac{1}{2\times5}+\frac{1}{5\times8}+\cdots+\frac{1}{14\times17}$
$$=\frac{1}{3}\left[\left(\frac{1}{2}-\frac{1}{5}\right)+\left(\frac{1}{5}-\frac{1}{8}\right)+\cdots+\left(\frac{1}{14}-\frac{1}{17}\right)\right]$$
$$=\frac{1}{3}\left(\frac{1}{2}-\frac{1}{17}\right)=\frac{5}{34}.$$

6. Let $A=\frac{1}{3}+\frac{1}{4}+\cdots+\frac{1}{2009}$, $B=\frac{1}{2}+\frac{1}{3}+\cdots+\frac{1}{2008}$, then the expression

becomes

$$A(1+B)-(1+A)B=A-B=\frac{1}{2009}-\frac{1}{2}=-\frac{2007}{4018}.$$

7. $\frac{25}{26}$. The nth term of the sum is

$$\frac{1}{1+2+\cdots+(n+1)}$$
$$=\frac{1}{\frac{(n+1)(n+2)}{2}}=\frac{2}{(n+1)(n+2)}=2\left(\frac{1}{n+1}-\frac{1}{n+2}\right).$$

Therefore

$$\frac{1}{1+2}+\frac{1}{1+2+3}+\cdots+\frac{1}{1+2+\cdots+51}$$
$$=2\left[\left(\frac{1}{2}-\frac{1}{3}\right)+\cdots+\left(\frac{1}{51}-\frac{1}{52}\right)\right]=2\left[\frac{1}{2}-\frac{1}{52}\right]=\frac{25}{26}.$$

8. $\frac{n(n+1)}{2}$. For any positive integer k we have

$$\frac{1}{k}+\frac{2}{k}+\cdots+\frac{k}{k}+\frac{k-1}{k}+\cdots+\frac{1}{k}=\frac{2\cdot\frac{(k-1)k}{2}+k}{k}=k,$$

the given sum becomes $1+2+\cdots+n=\frac{n(n+1)}{2}$.

9. 2019045. By using the formula $a^2-b^2=(a-b)(a+b)$,

$$1^2-2^2+3^2-4^2+\cdots+2007^2-2008^2+2009^2$$
$$=1+(3^2-2^2)+(5^2-4^2)+\cdots+(2009^2-2008^2)$$
$$=1+2+3+4+5+\cdots+2008+2009=\frac{2009\times 2010}{2}$$
$$=2019045.$$

10. 2222222184.

$$11+192+1993+19994+\cdots+199999998+1999999999$$
$$=20+200+2000+\cdots+2000000000-(9+8+\cdots+1)$$
$$=2222222220-\frac{9\times 10}{2}=2222222175.$$

Testing Questions (1-B)

1. By partitioning the integer and fractional part of each term,

$$\begin{aligned}
&\frac{3^2+1}{3^2-1}+\frac{5^2+1}{5^2-1}+\frac{7^2+1}{7^2-1}+\cdots+\frac{99^2+1}{99^2-1}\\
&=\left(1+\frac{2}{3^2-1}\right)+\left(1+\frac{2}{5^2-1}\right)+\cdots+\left(1+\frac{2}{99^2-1}\right)\\
&=49+\frac{1}{3-1}-\frac{1}{3+1}+\cdots+\frac{1}{99-1}-\frac{1}{99+1}\\
&=49+\frac{1}{2}-\frac{1}{100}=49\frac{49}{100}.
\end{aligned}$$

2. Let the sum be S. For any positive integer $n \geq 2$,

$$\begin{aligned}
&\frac{n}{[1+2+\cdots+(n-1)][1+2+\cdots+n]}=\frac{n}{\frac{(n-1)n}{2}\cdot\frac{n(n+1)}{2}}\\
&=\frac{4}{(n-1)n(n+1)}=2\left(\frac{1}{(n-1)n}-\frac{1}{n(n+1)}\right),
\end{aligned}$$

therefore

$$\begin{aligned}
S &= 1-\left(\frac{2}{1\cdot(1+2)}+\frac{3}{(1+2)(1+2+3)}+\cdots\right.\\
&\qquad\left.+\frac{100}{(1+2+\cdots+99)(1+2+\cdots+100)}\right)\\
&= 1-2\left[\left(\frac{1}{1\cdot 2}-\frac{1}{2\cdot 3}\right)+\left(\frac{1}{2\cdot 3}-\frac{1}{3\cdot 4}\right)+\cdots\right.\\
&\qquad\left.+\left(\frac{1}{99\cdot 100}-\frac{1}{100\cdot 101}\right)\right]\\
&= 1-2\left(\frac{1}{2}-\frac{1}{10100}\right)=\frac{1}{5050}.
\end{aligned}$$

Thus, the difference of the denominator and the numerator is 5049.

3. For any positive integer n,

$$\frac{1}{n(n+1)(n+2)}=\frac{1}{2}\left[\frac{1}{n(n+1)}-\frac{1}{(n+1)(n+2)}\right].$$

Therefore

$$\begin{aligned}
&\frac{1}{1\times2\times3}+\frac{1}{2\times3\times4}+\cdots+\frac{1}{100\times101\times102}\\
&=\frac{1}{2}\left[\left(\frac{1}{1\cdot2}-\frac{1}{2\cdot3}\right)+\cdots+\left(\frac{1}{100\cdot101}-\frac{1}{101\cdot102}\right)\right]\\
&=\frac{1}{2}\left[\frac{1}{2}-\frac{1}{101\times102}\right]=\frac{2575}{10302}.
\end{aligned}$$

4. For any positive integer n,

$$1+n^2+n^4=(n^2+1)^2-n^2=(n^2-n+1)(n^2+n+1),$$

so that

$$\begin{aligned}
\frac{n}{1+n^2+n^4} &= \frac{n}{(n^2-n+1)(n^2+n+1)}\\
&= \frac{1}{2}\left[\frac{1}{n(n-1)+1}-\frac{1}{n(n+1)+1}\right].
\end{aligned}$$

Therefore

$$\begin{aligned}
&\frac{1}{1+1^2+1^4}+\frac{2}{1+2^2+2^4}+\frac{3}{1+3^2+3^4}+\cdots+\frac{50}{1+50^2+50^4}\\
&=\frac{1}{2}\left[\left(\frac{1}{1}-\frac{1}{3}\right)+\left(\frac{1}{3}-\frac{1}{7}\right)+\cdots+\left(\frac{1}{50\cdot49+1}-\frac{1}{50\cdot51+1}\right)\right]\\
&=\frac{1}{2}\left[1-\frac{1}{50\cdot51+1}\right]=\frac{1}{2}\cdot\frac{2550}{2551}=\frac{1275}{2551}.
\end{aligned}$$

5. For each positive integer n,

$$\begin{aligned}
&\frac{n^2}{n^2-10n+50}+\frac{(10-n)^2}{(10-n)^2-10(10-n)+50}\\
&=\frac{2n^2}{n^2+(10-n)^2}+\frac{2(10-n)^2}{(10-n)^2+n^2}=2,
\end{aligned}$$

therefore

$$\begin{aligned}
&\frac{1^2}{1^2-10+50}+\frac{2^2}{2^2-20+50}+\cdots+\frac{9^2}{9^2-90+50}\\
&=\left(\frac{1^2}{1^2-10+50}+\frac{9^2}{9^2-90+50}\right)+\left(\frac{2^2}{2^2-20+50}+\frac{8^2}{8^2-80+50}\right)
\end{aligned}$$

$$+\left(\frac{3^2}{3^2-30+50}+\frac{7^2}{7^2-70+50}\right)+\left(\frac{4^2}{4^2-40+50}+\frac{6^2}{6^2-60+50}\right)$$

$$+\frac{5^2}{5^2-50+50}=2\times 4+1=9.$$

Solutions to Testing Questions 2

Testing Questions (2-A)

1. (B) and (C) are not monomials.
2. (D). For example $x^4+1+(-x^4+2)=3$ and $x^4+(-x^2)=x^4-x^2$.
3. Let the other polynomial be $P(x)$. From

$$P(x)+2x^2+x+1=P(x)-(2x^2+x+1)+2(2x^2+x+1)$$
$$=5x^2-2x+4+4x^2+2x+2=9x^2+6.$$

Thus, the sum is $9x^2+6$.

4. From the given conditions we have $b=m-1=n, c=2n-1=m$ and $0.75-0.5=1.25a$, therefore $2n-1=1+n$ i.e. $n=2, m=3$ and $a=0.2$, so that $b=2, c=3$. Thus, $abc=1.2$.
5. (C). The term with a greatest degree in the product is $x^5\cdot x=x^6$.
6. Since

$$\begin{aligned}2^8+2^{10}+2^n &= [(2^4)^2+2(2^4)(2^5)+(2^5)^2]+2^n-2^{10}\\ &= (2^4+2^5)^2+2^n-2^{10},\end{aligned}$$

$n=10$ is a solution.

7. From $3x^2+x=1$ we have $3x^2+x-1=0$. then

$$6x^3-x^2-3x+2010=2x(3x^2+x-1)-(3x^2+x-1)+2009=2009.$$

Thus, the value of the given expression is 2009.

8. (C). When $a+b+c\neq 0$, from the given equalities we have

$$a=(b+c)x,\quad b=(a+c)x,\quad c=(a+b)x.$$

By adding them up, we obtain $2(a+b+c)x = a+b+c$, so $x = \frac{1}{2}$.

Since $b+c \neq 0, a+c \neq 0, a+b \neq 0$, if one of a, b, c is 0, then, from the given equalities, the other two are zeros also, a contradiction. Therefore $abc \neq 0$.

When $a+b+c=0$, then $b+c=-a, a+c=-b, a+b=-c$, so

$$x = \frac{a}{b+c} = \frac{b}{a+c} = \frac{c}{a+b} = -1.$$

Thus, the answer is (C).

9. From the given equality we know that $xy \neq 0$. From $\frac{1}{x} - \frac{1}{y} = 4$ we have $y - x = 4xy$, therefore

$$\frac{2x+4xy-2y}{x-y-2xy} = \frac{4xy-2(y-x)}{-2xy-(y-x)} = \frac{4xy-8xy}{-2xy-4xy} = \frac{2}{3}.$$

Testing Questions (2-B)

1.

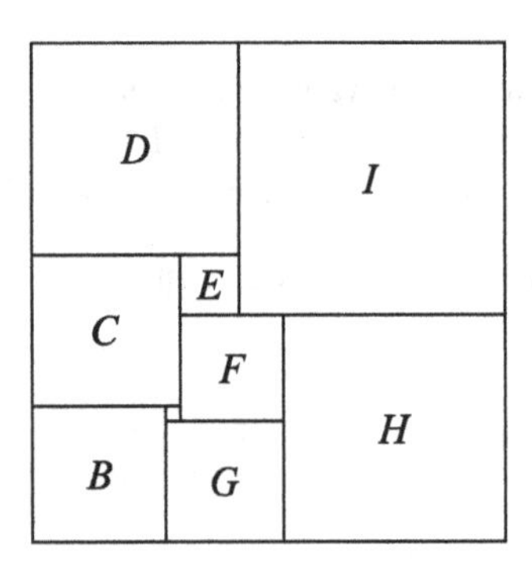

Let A be the smallest square which has sides of length $a = 1$, B has sides of length b, etc.

From the diagram we see that $a < b, a < c < d < i, a < f < g < h$; also $e < f < g < b < h$, and $e < c < d < i$. Hence E is the next smallest square.

Now $b = a+g, c = a+b = 2a+g, f = g-a$, and $c+a = f+e$. So

$$e = c+a-f = c+a-(g-a) = (2a+g)+a-g+a = 4a = 4.$$

Since $d + i = 2d + e = 2c + 3e = 2c + 12a$ and

$$b + g + h = a + 3g + f = 4a + 4f = 4a + 4(c + a - e) = 4c - 8a,$$

we have $2c + 12a = 4c - 8a$, so $c = 10a$. Hence $f = 7a, g = 8a, d = 14a, i = 18a, h = 15a$, and

$$\text{Area of the rectangle} = 33a \times 32a = 1056a^2 = 1056.$$

2. From $P(-7) = a(-7)^7 + b(-7)^3 + c(-7) - 5 = 7$, we have

$$-[a(7)^7 + b(7)^3 + c(7)] = 12,$$

therefore $P(7) = a(7)^7 + b(7)^3 + c(7) - 5 = -12 - 5 = -17$.

3. From $\frac{1}{a} + \frac{1}{b} + \frac{1}{c} = \frac{1}{a+b+c}$,

$$\frac{bc + ca + ab}{abc} = \frac{1}{a+b+c},$$
$$(a + b + c)(bc + ca + ab) = abc,$$
$$(a + b)(bc + ca + ab) + (bc^2 + ac^2) + abc - abc = 0,$$
$$(a + b)(bc + ca + ab) + (a + b)c^2 = 0,$$
$$(a + b)[(bc + c^2) + (ca + ab)] = 0,$$
$$(a + b)[(b + c)c + (c + b)a] = 0,$$
$$(a + b)(b + c)(c + a) = 0.$$

Thus, $a + b = 0$ or $b + c = 0$ or $c + a = 0$.

4. From the given equalities we have $x^2y^2z^2 = abc$, when it is divided by $y^2z^2 = c^2$, we obtain

$$x^2 = \frac{abc}{c^2} = \frac{ab}{c}.$$

Similarly, we have $y^2 = \frac{ca}{b}$, $z^2 = \frac{bc}{a}$. Thus,

$$x^2 + y^2 + z^2 = \frac{ab}{c} + \frac{ca}{b} + \frac{bc}{a} = \frac{(ab)^2 + (ca)^2 + (bc)^2}{abc}.$$

5. From $a^4 + a^3 + a^2 + a + 1 = 0$ we find that $a \neq 1$, i.e. $a - 1 \neq 0$. From

$$a(a^4 + a^3 + a^2 + a + 1) - (a^4 + a^3 + a^2 + a + 1)$$
$$= a^5 + a^4 + a^3 + a^2 + a - a^4 - a^3 - a^2 - a - 1 = a^5 - 1,$$

we find that $a^5 = 1$. Therefore

$$a^{2000} + a^{2010} + 1 = (a^5)^{400} + (a^5)^{402} + 1 = 1 + 1 + 1 = 3.$$

6. Let $x = 1$ in the given equality, we obtain

$$(-1)^n = a_{2n} + a_{2n-1} + \cdots + a_1 + a_0.$$

Let $x = -1$ in the given equality, we obtain

$$1 = a_{2n} - a_{2n-1} + a_{2n-2} - a_{2n-3} + \cdots - a_1 + a_0.$$

By adding them up, we have

$$1 + (-1)^n = 2(a_0 + a_2 + a_4 + \cdots + a_{2n}),$$

i.e.

$$a_0 + a_2 + a_4 + \cdots + a_{2n} = \frac{1 + (-1)^n}{2} = \begin{cases} 1 & \text{for even } n, \\ 0 & \text{for odd } n. \end{cases}$$

Solutions to Testing Questions 3

Testing Questions (3-A)

1. (D).

2. From $x = \frac{12}{k}$ which is a positive integer, k is also positive integer and k is a divisor of 12. So

$$k = 1,\ 2,\ 3,\ 4,\ 6,\ 12.$$

The number of possible values of k is 6.

3. From $\dfrac{3}{x+2} < \dfrac{13}{12} < \dfrac{3}{x}$, it follows that

$$13x < 36 < 13(x+2),$$
$$x < 3, \quad \text{i.e. } x = 1 \text{ or } 2.$$

By checking, $x = 1$ does not satisfy the original equation, and $x = 2$ satisfies the given equation. Thus, the answer is (B).

4. Substituting 4 into the given equation as x, it follows that

$$3a - 4 = 2 + 3 = 5 \Longrightarrow a = 3.$$

Thus $(-a)^2 - 2a = 9 - 6 = 3$.

5. From the given equation we have

$$\frac{n(x-n)-m(x-m)}{mn}=\frac{m}{n},$$
$$(n-m)x-n^2+m^2=m^2,$$
$$(n-m)x=n^2.$$

When $n \neq m$, we have $x=\dfrac{n^2}{n-m}$; when $n=m$, no solution.

6. When $a+b=0$, any real number is a solution of the equation.
When $a+b \neq 0$, from $4ax-(a+b)=0$ we have $4ax=a+b$, so the equation has no solution if $a=0$, and $x=\dfrac{a+b}{4a}$ if $a \neq 0$.

7. From $\frac{1}{3}m(-2)=5(-2)+(-2)^2$, we obtain $m=9$. Therefore

$$(m^2-11m+17)^{2007}=(81-99+17)^{2007}=(-1)^{2007}=-1.$$

8. From the given equation we have $m^2x+1=mx+m$, therefore

$$m(m-1)x=m-1.$$

(i) When $m \neq 1$ and $m \neq 0$, then $x=\dfrac{m-1}{m(m-1)}=\dfrac{1}{m}$;

(ii) When $m=1$, the equation becomes $0 \cdot x=0$, so any real number is a solution for x;

(iii) When $m=0$, then the equation becomes $0 \cdot x=-1$, no solution for x.

9. After arranging the terms of the given equation, we obtain $(k^2-2k)x=k^2-5k$, so

$$(k-2)x=k-5 \quad \text{or} \quad (k-2)^2x=(k-5)(k-2).$$
$$\therefore (k-5)(k-2)>0 \text{ i.e. } (k-5) \text{ and } (k-2) \text{ have same sign.}$$

Thus, $k>5$ or $0<k<2$.

10. From the given equation we have $(2a-3)x = a-3$. The equation has no solution for x means

$$2a-3=0 \quad \text{and} \quad a-3 \neq 0,$$

so $a = \frac{3}{2}$.

Testing Questions (3-B)

1. From the first equation we have

$$3x - 6(x + \frac{a}{3}) = 2x,$$

$$3x - 6x - 2a = 2x,$$

$$5x = -2a, \ \therefore x = -\frac{2a}{5}.$$

From the second equation we have

$$2(3x+a) - (1+4x) = 0$$

$$2x = 1-2a, \ \therefore x = \frac{1-2a}{2}.$$

Hence we have equation in a: $-\frac{2a}{5} = \frac{1-2a}{2}$. By solving it, we have

$$-4a = 5(1-2a) = 5 - 10a \Longrightarrow 6a = 5 \Longrightarrow a = \frac{5}{6}.$$

Thus,

$$x = -\frac{2a}{5} = -\frac{2}{5} \cdot \frac{5}{6} = -\frac{1}{3}.$$

2. From $abc = 1$, the given equation can be changed in the form

$$\frac{2abcx}{ab \cdot bc + a \cdot bc + bc} + \frac{2bx}{bc+b+1} + \frac{2bcx}{ca \cdot b + c \cdot b + b} = 1,$$

$$\frac{2x}{b+1+bc} + \frac{2bx}{bc+b+1} + \frac{2bcx}{1+bc+b} = 1,$$

$$\frac{2(1+b+bc)x}{bc+b+1} = 1, \ \therefore x = \frac{1}{2}.$$

3. Let the positive integer solution be x. From the given equation we have

$$m = \left(\frac{8}{3} - \frac{9}{4}\right)x - 123 = \frac{5}{12}x - 123.$$

Since m is a positive integer, $\frac{5}{12}x - 123 \geq 1$, so $x = 12k$ and

$$k = \frac{x}{12} \geq \frac{124}{5} = 24.8,$$

i.e. the minimum value of k is 25. Thus, $x = 300, m_{\min} = 5k - 123 = 2$.

4. From $3[4x - (2x - 6)] = 11x + 8$ we have

$$3(2x + 6) = 11x + 8 \Longrightarrow 6x + 18 = 11x + 8 \Longrightarrow 5x = 10 \Longrightarrow x = 2.$$

Therefore there are infinitely many required equations, say $\frac{1}{4}x - \frac{1}{2} = 0$ satisfies the requirement.

5. For $n = 1, 2, \ldots, 2008$,

$$a_{n+1} = \frac{1}{1 + \frac{1}{a_n}} \Longrightarrow a_{n+1} = \frac{a_n}{a_n + 1} \Longrightarrow a_{n+1}a_n + a_{n+1} = a_n$$
$$\Longrightarrow a_n a_{n+1} = a_n - a_{n+1},$$

therefore

$$\begin{aligned} &a_1a_2 + a_2a_3 + a_3a_4 + \cdots + a_{2008}a_{2009} \\ &= (a_1 - a_2) + (a_2 - a_3) + \cdots + (a_{2008} - a_{2009}) \\ &= a_1 - a_{2009}. \end{aligned}$$

On the other hand, From the formula $a_{n+1} = \frac{a_n}{a_n + 1}$ we find that

$$a_2 = \frac{1}{2}, \quad a_3 = \frac{1}{3}, \quad a_4 = \frac{1}{4}.$$

Assuming $a_{n-1} = \frac{1}{n-1}$, we obtain $a_n = \frac{\frac{1}{n-1}}{\frac{1}{n-1} + 1} = \frac{1}{n}$, therefore $a_{2009} = \frac{1}{2009}$, and

$$a_1a_2 + a_2a_3 + a_3a_4 + \cdots + a_{2008}a_{2009} = 1 - \frac{1}{2009} = \frac{2008}{2009}.$$

Solutions to Test questions 4

Testing Questions (4-A)

1. By substituting the solution (2, 1) into equations, we obtain

$$\begin{cases} 2a + b = 7, \\ 2b + c = 5. \end{cases}$$

After eliminating b, we obtain $4a - c = 9$. The answer is (C).

2. From the first equation of the first system we get $y = 3x - 5$. By substituting it into the last equation of the second system, it follows that $2x + 3(3x - 5)) = -4$, so $x = 1, y = -2$. Then, the second equation of the first system yields $z = 0$. Thus, from the second equation of the second system, $c = 4$. By solving the system

$$4a - 10b = -22, \quad a + 2b = 8,$$

the solution for a and b is obtained: $a = 2, b = 3$. Thus, the answer is (A).

3. The given system can be expressed in the form

$$\begin{cases} kx - y = -\frac{1}{3}, \\ 6x + 3y = 1. \end{cases}$$

(1) When $\frac{k}{6} \neq \frac{-1}{3}$, i.e. $k \neq -2$, the system has unique solution

$$x = 0 \quad y = -\frac{1}{3}.$$

(2) When $\frac{k}{6} = -\frac{1}{3}$, the system has infinitely many solutions.

(3) Thus, its impossible that the system has no solution.

4. Since $abc \neq 0$, by writing down the given equations in the form

$$\frac{a+b}{ab} = \frac{1}{2}, \quad \frac{a+c}{ac} = \frac{1}{5}, \quad \frac{b+c}{bc} = \frac{1}{4},$$

we obtain the new system satisfied by a, b, c:

$$\frac{1}{a}+\frac{1}{b}=\frac{1}{2}, \tag{15.1}$$
$$\frac{1}{a}+\frac{1}{c}=\frac{1}{5}, \tag{15.2}$$
$$\frac{1}{b}+\frac{1}{c}=\frac{1}{4}. \tag{15.3}$$

By $\frac{1}{2}((15.1)+(15.2)+(15.3))$, it follows that

$$\frac{1}{a}+\frac{1}{b}+\frac{1}{c}=\frac{19}{40}. \tag{15.4}$$

By (15.4) − (15.1), then $\frac{1}{c}=-\frac{1}{40}$, i.e. $c=-40$.

By (15.4) − (15.2), then $\frac{1}{b}=\frac{11}{40}$, i.e. $b=\frac{40}{11}$.

By (15.4) − (15.3), then $\frac{1}{a}=\frac{9}{40}$, i.e. $a=\frac{40}{9}$.

5. Let the given equations be equations (15.5), (15.6), (15.7), respectively.

$$x-y-z = 5 \tag{15.5}$$
$$y-z-x = 1 \tag{15.6}$$
$$z-x-y = -15. \tag{15.7}$$

By (15.5) + (15.6) + (15.7),

$$x+y+z=9. \tag{15.8}$$

By (15.6) + (15.7), it follows that $2x=14$, i.e. $x=7$. Similarly, by (15.6) + (15.8) and (15.7) + (15.8) respectively we obtain

$$y=5, \qquad z=-3.$$

Thus, the solution is $x=7,\ y=5,\ z=-3$.

6. Let

$$x-y+z = 1 \tag{15.9}$$
$$y-z+u = 2 \tag{15.10}$$
$$z-u+v = 3 \tag{15.11}$$
$$u-v+x = 4 \tag{15.12}$$
$$v-x+y = 5. \tag{15.13}$$

By (15.9) + (15.10) + (15.11) + (15.12) + (15.13), it follows that

$$x + y + z + u + v = 15. \tag{15.14}$$

By (15.9) + (15.10), (15.10) + (15.11), (15.11) + (15.12), (15.12) + (15.13),

(15.13) + (15.9) respectively, we obtain

$$\begin{aligned} x + u &= 3 & (15.15)\\ y + v &= 5 & (15.16)\\ z + x &= 7 & (15.17)\\ u + y &= 9 & (15.18)\\ v + z &= 6. & (15.19) \end{aligned}$$

By (15.15) + (15.16) + (15.17) − (15.14), (15.16) + (15.17) + (15.18) − (15.14), (15.17) + (15.18) + (15.19) − (15.14), (15.18) + (15.19) + (15.15) − (15.14), (15.19) + (15.15) + (15.16) − (15.14) respectively, it follows that

$$x = 0, \quad y = 6, \quad z = 7, \quad u = 3, \quad v = -1.$$

7. By (4.25) − (4.26) to eliminate x, it follows that
$$\frac{8}{y} + \frac{8}{z} = 0, \text{ i.e. } \frac{y}{z} = -1.$$

   $3 \times (4.25) + (4.26)$ eliminates y and yields $\frac{4}{x} + \frac{4}{z} = 0$, i.e. $\frac{z}{x} = -1$.

   $5 \times (4.25) + 3 \times (4.26)$ eliminates z and yields $\frac{8}{x} - \frac{8}{y} = 0$, i.e. $\frac{x}{y} = 1$.

   Thus, $\frac{x}{y} + \frac{y}{z} + \frac{z}{x} = 1 - 1 - 1 = -1$.

8. By adding the two equations, we obtain $(m + 3)x = 10$, so

$$x = \frac{10}{m+3}, \qquad y = \frac{3}{2}x = \frac{15}{m+3}.$$

   Since $m + 3 \mid 10$ and $m + 3 \mid 15$, so $m + 3 \mid 5$. Thus, $m = \pm 2$ and $m^2 = 4$.

9. Let $S = a + b + c + d + e + f$. Then

a	b	6
c	d	e
f	7	2

$f+d+6=f+7+2 \Longrightarrow d=3.$
$a+3+2=f+7+2 \Longrightarrow a=f+4.$
$e+8=f+9 \Longrightarrow e=f+1$
$a+f=3+f+1 \Longrightarrow f=0.$
$\therefore a=4, e=1, c=9-a=5,$
$b=9-4-6=-1.$ Thus

$$a+b+c+d+e+f=3\times 9-6-7-2=12.$$

10. By (4.29) – (4.28) – (4.27), we obtain $-4z=-12$, so $z=3$. Substituting it into the equations (4.27), (4.29), and (4.30), it follows that

$$x+y+u=7, \qquad (15.20)$$

$$3x+2y+4u=24, \qquad (15.21)$$

$$4x+3y+2u=19. \qquad (15.22)$$

By (15.22) – (15.20) – (15.21), we obtain $-3u=-12$, so $u=4$. Substituting it into (15.20) and (15.21), it follows that

$$x+y=3, \qquad 3x+2y=8.$$

By solving it, we obtain $x=2, y=1$. Thus, $x=2, y=1, z=3, u=4$.

Testing Questions (4-B)

1. From that the system has no solution we find that $\frac{3}{2}=\frac{m}{n}\neq\frac{7}{4}$. Therefore $m=\frac{3n}{2}$, and n is even satisfying $-9\cdot\frac{2}{3}\le n\le 9\cdot\frac{2}{3}$, i.e.

$$-6\le n\le 6.$$

So $n=-6,-4,-2,0,2,4,6$, and correspondingly, $m=-9,-6,-3,0,3,6,9$. That is, (m,n) can be one of

$$(-9,-6),\ (-6,-4),\ (-3,-2),\ (0,0),\ (3,2),\ (6,4),\ (9,6).$$

2. Combine the left hand side of each equation, then let $x+y+z=t$, we have

$$xy+xz=2t, \qquad (15.23)$$

$$yz+xy=3t, \qquad (15.24)$$

$$zx+yz=4t. \qquad (15.25)$$

Besides, $\frac{1}{2}((15.23) + (15.24) + (15.25))$ yields

$$xy + yz + zx = \frac{9}{2}t. \qquad (15.26)$$

From (15.26) – (15.23), (15.26) – (15.24), (15.26) – (15.25), respectively, it follows that

$$xy = \frac{1}{2}t, \quad yz = \frac{5}{2}t, \quad zx = \frac{3}{2}t.$$

Since $x, y, z \neq 0$, $t \neq 0$. It's easy to see that $x : y : z = 3 : 5 : 15$, therefore

$$x = \frac{3}{23}t, \quad y = \frac{5}{23}t, \quad z = \frac{15}{23}t,$$

$$xy = \frac{15}{23^2}t^2 = \frac{1}{2}t \Longrightarrow t = \frac{23^2}{30}.$$

Thus, $x = \frac{23}{10}$, $y = \frac{23}{6}$, $z = \frac{23}{2}$. By checking, the triple satisfies the original equation, so it is the solution.

3. The system can be rewritten in the form

$$x(x + y + z) = 60, \qquad (15.27)$$

$$y(x + y + z) = 75, \qquad (15.28)$$

$$z(x + y + z) = 90. \qquad (15.29)$$

By adding them, we obtain $(x + y + z)^2 = 225$, i.e.

$$x + y + z = \pm 15. \qquad (15.30)$$

By substituting back (15.30) into (15.27), (15.28), (15.29) respectively, we obtain $x = 4, y = 5, z = 6$ or $x = -4, y = -5, z = -6$.

4. By $(4.31) + 2 \times (4.32)$, it follows that $x = \frac{56 - 3a}{5} = 11 + \frac{1 - 3a}{5}$ and by $2 \times (4.31) - (4.32)$, it follows that $5y = 4a - 13$, so $y = \frac{4a - 3}{5} - 2$.

$56 - 3a \geq 5$ and $4a - 13 \geq 5$ implies that $4 < a \leq 19$. From $5 \mid (1 - 3a)$, the units' digit of a may be 2 or 7. It's easy to see that when $a = 12$ the system has the solution $x = 4, y = 7$, and when $a = 17$ the system has the solution $x = 1, y = 11$. Thus, a can be 12 and 17.

5. By adding up (4.33), $\cdots$, (4.37) and then divided by 6, it follows that

$$x + y + z + u + v = 16. \qquad (15.31)$$

Use each of the given 5 equations minus (15.31), we obtain

$$x = 0, \quad y = 1, \quad z = 3, \quad u = 5, \quad v = 7.$$

Solutions to Testing Questions 5

Testing Question (5-A)

1. From $a^2 + b^2 + 8a - 14b + 65 = 0$,

$$(a^2 + 8a + 16) + (b^2 - 14b + 49) = 0,$$
$$(a + 4)^2 + (b - 7)^2 = 0.$$

Since $(a + 4)^2 \geq 0$ and $(b - 7)^2 \geq 0$ for any real numbers a and b, we obtain $a + 4 = 0$ and $b - 7 = 0$, i.e. $a = -4,\ b = 7$. Therefore

$$a^2 + ab + b^2 = (-4)^2 - (4)(7) + 7^2 = 37.$$

2. From $a - b = 2, b - c = 4$ we have $c - a = -6$. Thus

$$a^2 + b^2 + c^2 - ab - bc - ca$$
$$= \frac{1}{2}[(a-b)^2 + (b-c)^2 + (c-a)^2] = 2 + 8 + 18 = 28.$$

3.

$$(a^2 + b^2)(c^2 + d^2) = a^2c^2 + a^2d^2 + b^2c^2 + b^2d^2$$
$$= [(ac)^2 + 2abcd + (bd)^2] + [(ad)^2 - 2abcd + (bc)^2]$$
$$= (ac + bd)^2 + (ad - bc)^2.$$

4. From the given equality,

$$14a^2 + 14b^2 + 14c^2 = a^2 + 4b^2 + 9c^2 + 4ab + 6ac + 12bc,$$
$$13a^2 + 10b^2 + 5c^2 - 4ab - 6ac - 12bc = 0,$$
$$(4a^2 - 4ab + b^2) + (9b^2 - 12bc + 4c^2) + (9a^2 - 6ac + c^2) = 0,$$
$$(2a - b)^2 + (3b - 2c)^2 + (3a - c)^2 = 0.$$

Since any square is non-negative, we have $2a - b = 0,\ 3b - 2c = 0,\ 3a - c = 0$, therefore

$$b = 2a,\ c = 3a.$$

Thus, $a : b : c = 1 : 2 : 3$.

5. It is obvious that $x \neq 0$. Then

$$\frac{x}{x^2 + 3x + 1} = a \iff \frac{1}{\left(x + \frac{1}{x}\right) + 3} = a \iff x + \frac{1}{x} = \frac{1}{a} - 3.$$

Therefore

$$\frac{x^2}{x^4+3x^2+1} = \frac{1}{\left(x^2+\frac{1}{x^2}\right)+3} = \frac{1}{\left(x+\frac{1}{x}\right)^2+1}$$
$$= \frac{1}{\left(\frac{1-3a}{a}\right)^2+1} = \frac{a^2}{(1-3a)^2+a^2} = \frac{a^2}{10a^2-6a+1}.$$

6. First of all, we have $x^2 + \frac{1}{x^2} = a^2 - 2$. Then

$$x^6 + \frac{1}{x^6} = (x^2)^3 + \left(\frac{1}{x^2}\right)^3 = \left(x^2+\frac{1}{x^2}\right)\left(x^4+\frac{1}{x^4}-1\right)$$
$$= (a^2-2)[(a^2-2)^2-3] = (a^2-2)^3 - 3(a^2-2).$$

7. $0 = a^4 + b^4 + c^4 + d^4 - 4abcd$
$= (a^4 - 2a^2b^2 + b^4) + (c^4 - 2c^2d^2 + d^4) + 2(a^2b^2 - 2abcd + c^2d^2)$
$= (a^2 - b^2)^2 + (c^2 - d^2)^2 + 2(ab - cd)^2,$

therefore $a^2 = b^2, c^2 = d^2, ab = cd$, and they imply $a^2 = c^2$. Thus, the conclusion is proven.

8 From $a + b + c + d = 0$ we have $a + b = -(c + d)$. By taking power 3,

$$a^3 + 3a^2b + 3ab^2 + b^3 = -(c^3 + 3c^2d + 3cd^2 + c^3)$$
$$= -c^3 - 3c^2d - 3cd^2 - d^3,$$
$$a^3 + b^3 + c^3 + d^3 = -3a^2b - 3ab^2 - 3c^2d - 3cd^2$$
$$= 3ab[-(a+b)] + 3cd[-(c+d)],$$

$\therefore a^3 + b^3 + c^3 + d^3 = 3ab(c + d) + 3cd(a + b) = 3(abc + bcd + cda + dab)$.

9. Let $x - 2 = u, y - 2 = v, z - 2 = w$. Then $u^3 + v^3 + w^3 = 0$ and $u + v + w = 0$. From the identity

$$u^3 + v^3 + w^3 - 3uvw = (u + v + w)(u^2 + v^2 + w^2),$$

we have $-3uvw = 0$, i.e.

$$(x-2)(y-2)(z-2) = 0,$$

so $x - 2 = 0$ or $y - 2 = 0$ or $z - 2 = 0$, the conclusion is proven.

10 From

$$\begin{aligned}&(a+b+c)^3-a^3-b^3-c^3=[(a+b+c)^3-a^3]-(b^3+c^3)\\&=(b+c)[(a+b+c)^2+(a+b+c)a+a^2]-(b+c)(b^2-bc+c^2)\\&=(b+c)[3a^2+b^2+c^2+3ab+3ca+2bc-b^2-c^2+bc)\\&=3(b+c)(a^2+ab+bc+ca)=3(b+c)[(a^2+ab)+(bc+ca)]\\&=3(b+c)[a(a+b)+c(a+b)]=3(b+c)(c+a)(a+b).\end{aligned}$$

The given equation means that $(a+b+c)^3-a^3-b^3-c^3=0$, therefore

$$3(b+c)(c+a)(a+b)=0,$$

which implies $b+c=0$ or $c+a=0$ or $a+b=0$. In anyone of the three possible cases, we have the equality

$$a^{2n+1}+b^{2n+1}+c^{2n+1}=(a+b+c)^{2n+1}.$$

Testing Questions (5-B)

1. From

$$\begin{aligned}M&=3x^2-8xy+9y^2-4x+6y+13\\&=2(x^2-4xy+4y^2)+(x^2-4x+4)+(y^2+6y+9)\\&=2(x-2y)^2+(x-2)^2+(y+3)^2\ge 0.\end{aligned}$$

M is not negative. Further, the system $x-2y=0, x=2, y+3=0$ has no solution for (x,y), so M must be positive, i.e. the answer is (A).

2. From the given conditions we have $a-c=d-b$ and $a^2-c^2=d^2-b^2$, therefore

$$(a-c)(a+c)=(d-b)(d+b).$$

If $a-c=0=d-b$, the conclusion is true obviously. If $a-c=d-b\neq 0$, then $a+c=d+b$. Considering $a-c=d-b$, we obtain

$$a=d \text{ and } c=b,$$

so the conclusion holds also.

3. From $a+b+c=0$,

$$\begin{aligned}&2(a^4+b^4+c^4)-(a^2+b^2+c^2)^2=a^4+b^4+c^4-2a^2b^2-2b^2c^2-2c^2a^2\\&=(a^2-b^2-c^2)^2-4b^2c^2=(a^2-b^2-c^2)^2-(2bc)^2\\&=(a^2-b^2-c^2+2bc)(a^2-b^2-c^2-2bc)\\&=[(a^2-(b-c)^2][(a^2-(b+c)^2]\\&=(a-b+c)(a+b-c)(a-b-c)(a+b+c)=0.\end{aligned}$$

4. From $(a^3+b^3)(a^4+b^4) = (a^7+b^7) + (ab)^3(a+b)$ we have

$$a^7+b^7 = (a^3+b^3)(a^4+b^4) - (ab)^3(a+b)$$
$$= (a+b)[(a+b)^2 - 3ab] \cdot [(a^2+b^2)^2 - 2(ab)^2] - (ab)^3(a+b).$$

Therefore it suffices to find the value of ab. Since

$$ab = \frac{1}{2}[(a+b)^2 - (a^2+b^2)] = \frac{1}{2}[1-2] = -\frac{1}{2},$$

we have

$$a^7+b^7 = \left[1 - 3\left(-\frac{1}{2}\right)\right] \cdot \left[2^2 - 2\left(-\frac{1}{2}\right)^2\right] + \left(\frac{1}{2}\right)^3 = \frac{5}{2} \cdot \frac{7}{2} + \frac{1}{8} = \frac{71}{8}.$$

5. Let $a+b=x$. Then $a^3+b^3+3ab=1$ implies

$$(a+b)[(a+b)^2 - 3ab] + 3ab - 1 = 0,$$
$$x^3 - 3abx + 3ab - 1 = 0.$$

Since

$$x^3 - 3abx + 3ab - 1 = (x-1)(x^2+x+1) - 3ab(x-1)$$
$$= (x-1)(x^2+x+1-3ab),$$

we obtain $(x-1)(x^2+x+1-3ab) = 0$.

When $x-1=0$, we have $a+b=1$. When $x^2+x+1-3ab=0$, then $(a+b)^2+a+b+1-3ab=0$, i.e. $a^2+b^2-ab+a+b+1=0$. By completing squares, we have

$$\frac{1}{2}[(a-b)^2 + (a+1)^2 + (b+1)^2] = 0,$$

hence $a=b=-1$, so $a+b=-2$. By checking, $a+b=1$ or $a=b=-1$ satisfy the original equation. Thus, $a+b=1$ or $a+b=-2$.

Solutions to Testing Questions (6)

Testing Questions (6-A)

1. Factorizations:

(i) $x^9 + 7x^6y^3 + 7x^3y^6 + y^9 = [(x^3)^3 + (y^3)^3] + 7x^3y^3(x^3 + y^3)$
$= (x^3 + y^3)(x^6 - x^3y^3 + y^6) + 7x^3y^3(x^3 + y^3)$
$= (x^3 + y^3)(x^4 + 6x^3y^3 + y^4).$

(ii) $4x^2 + y^2 + 9z^2 - 6yz + 12zx - 4xy$
$= (2x)^2 + (-y)^2 + (3z)^2 + 2(-y)(3z) + 2(3z)(2x) + 2(2x)(-y)$
$= (2x - y + 3z)^2.$

(iii) Let $y = x + 2$, then
$(x^2 - 1)(x + 3)(x + 5) + 16 = (x - 1)(x + 1)(x + 3)(x + 5) + 16$
$= (y - 3)(y - 1)(y + 1)(y + 3) + 16$
$= (y^2 - 1)(y^2 - 9) + 16 = y^4 - 10y^2 + 25$
$= (y^2 - 5)^2 = [(x + 2)^2 - 5]^2 = (x^2 + 4x - 1)^2.$

(iv) $(2x^2 - 4x + 1)^2 - 14x^2 + 28x + 3$
$= (2x^2 - 4x + 1)^2 - 7(2x^2 - 4x + 1) + 10$
$= (2x^2 - 4x + 1 - 2)(2x^2 - 4x + 1 - 5)$
$= (2x^2 - 4x - 1)(2x^2 - 4x - 4) = 2(2x^2 - 4x - 1)(x^2 - 2x - 2).$

(v) $x^3 - 3x^2 + (a + 2)x - 2a = (x^3 - 2x^2) - [x^2 - 2(a + 1)x + 4a]$
$= x^2(x - 2) - (x - 2a)(x - 2) = (x - 2)(x^2 - x + 2a).$

(vi) $x^{11} + x^{10} + \cdots + x^2 + x + 1 = \dfrac{x^{12} - 1}{x - 1} = \dfrac{(x^6 - 1)(x^6 + 1)}{x - 1}$

$$= \frac{(x^2 - 1)(x^4 + x^2 + 1)(x^6 + 1)}{x - 1}$$
$$= (x + 1)[(x^2 + 1)^2 - x^2](x^6 + 1)$$
$$= (x + 1)(x^2 - x + 1)(x^2 + x + 1)(x^2 + 1)(x^4 - x^2 + 1).$$

2. Factorizations:

(i) $x^4 - 2(a^2 + b^2)x^2 + (a^2 - b^2)^2$
$= [x^2 - (a^2 + b^2)]^2 - (a^2 + b^2)^2 + (a^2 - b^2)^2$
$= [x^2 - (a^2 + b^2)]^2 - 4a^2b^2$
$= (x^2 - a^2 - b^2 - 2ab)(x^2 - a^2 - b^2 + 2ab)$
$= [x^2 - (a + b)^2][x^2 - (a - b)^2]$
$= (x - a - b)(x + a + b)(x - a + b)(x + a - b).$

(ii) $(ab + 1)(a + 1)(b + 1) + ab = (ab + 1)(ab + a + b + 1) + ab$
$= (ab + 1)(ab + b + 1) + (ab + 1)a + ab$
$= (ab + 1)(ab + b + 1) + a(ab + b + 1)$
$= (ab + b + 1)(ab + a + 1).$

3. From

$$81^6 - 9\cdot 27^7 - 9^{11} = 9^{12} - 3\cdot 9^{11} - 9^{11} = 9^{11}(9 - 3 - 1) = 5\cdot 9^{11} = 45\cdot 9^{10},$$

the expression has a factor 45.

4. $\underbrace{44\cdots44}_{2n\text{ digits}} - \underbrace{88\cdots88}_{n\text{ digits}}$

$$= \underbrace{44\cdots44}_{n\text{ digits}}\cdot 10^n + \underbrace{44\cdots44}_{n\text{ digits}} - \underbrace{88\cdots88}_{n\text{ digits}} = \underbrace{44\cdots44}_{n\text{ digits}}\cdot 10^n - \underbrace{44\cdots44}_{n\text{ digits}}$$
$$= \underbrace{44\cdots44}_{n\text{ digits}}(10^n - 1) = \underbrace{44\cdots44}_{n\text{ digits}}\cdot\underbrace{99\cdots99}_{n\text{ digits}}$$
$$= (\underbrace{11\cdots11}_{n\text{ digits}})^2\cdot 36 = (\underbrace{66\cdots66}_{n\text{ digits}})^2.$$

5. Factorizations:

(i) Let $y = x^2 + x - 1$, then
$(x^2 + x - 1)^2 + x^2 + x - 3 = y^2 + y - 2 = (y - 2)(y + 1)$
$= (x^2 + x - 3)(x^2 + x) = x(x + 1)(x^2 + x - 3)$.

(ii) Let $u = x - y,\ v = y - x - 2,\ w = 2$, then $u + v + w = 0$,
$(x - y)^3 + (y - x - 2)^3 + 8 = u^3 + v^3 + w^3$
$= 3uvw + (u + v + w)(u^2 + v^2 + w^2 - uv - vw - wu)$
$= 3uvw = 3(x - y)(y - x - 2)(2) = -6(x - y)(x - y + 2)$.

(iii) Let $y = 6x + 5$, then

$$(6x + 5)^2(3x + 2)(x + 1) - 6 = \frac{1}{12}y^2(y - 1)(y + 1) - 6$$
$$= \frac{1}{12}y^2(y^2 - 1) - 6 = \frac{1}{12}(y^4 - y^2 - 72)$$
$$= \frac{1}{12}(y^2 - 9)(y^2 + 8) = \frac{1}{12}[(6x + 5)^2 - 3^2][(6x + 5)^2 + 8]$$
$$= \frac{1}{12}(6x + 8)(6x + 2)(36x^2 + 60x + 33)$$
$$= (3x + 4)(3x + 1)(12x^2 + 20x + 11).$$

(iv) Let $y = x^2 + 5x + 6$, then
$(x^2 + 5x + 6)(x^2 + 6x + 6) - 2x^2 = y(y + x) - 2x^2$
$= y^2 + xy - 2x^2 = (y + 2x)(y - x)$
$= (x^2 + 7x + 6)(x^2 + 4x + 6) = (x + 1)(x + 6)(x^2 + 4x + 6)$.

(v) Let $u = x^2 - 2x,\ v = x^2 - 4x + 2,\ w = -2(x^2 - 3x + 1)$,
then $u + v + w = 0$, so $u^3 + v^3 + w^3 = 3uvw$
$= -6(x^2 - 4x)(x^2 - 4x + 2)(x^2 - 3x + 1)$.

(vi) Let $t = a + b + c$. Then
$a^3 + b^3 + c^3 + (a + b)(b + c)(c + a) - 2abc$
$= (a^3 + b^3 + c^3 - 3abc) + (t - c)(t - a)(t - b) + abc$
$= t(a^2 + b^2 + c^2 - ab - bc - ca) + t^3 - (a + b + c)t^2$
$+(ca + bc + ab)t = (a^2 + b^2 + c^2)t = (a + b + c)(a^2 + b^2 + c^2)$.

6. (i) From $x^2 - xy - 2y^2 = (x-2y)(x+y)$, we let $x^2 - xy - 2y^2 + 8x + ay - 9 = (x - 2y + b)(x + y + c)$, the constants are to be determined below. Then

$$x^2-xy-2y^2+8x+ay-9 = x^2-xy-2y^2+(b+c)x+(b-2c)y+bc.$$

By the comparison of coefficients, the system of equations in b, c is obtained:

$$b+c=8, \quad b-2c=a, \quad bc=-9.$$

Then the first and third equations yield the solutions for (b, c):

$$(b,c)=(9,-1) \quad \text{or} \quad (b,c)=(-1,9).$$

If $b=9, c=-1$, then $a=b-2c=11$;

If $b=-1, c=9$, then $a=-1-18=-19$. Thus, $a=11$ or -19.

(ii) Let $x^4 - x^3 + 4x^2 + 3x + 5 = (x^2 + ax + b)(x^2 + cx + d)$. By expansion,

$$(x^2+ax+b)(x^2+cx+d) = x^4+(a+c)x^3+(ac+b+d)x^2+(ad+bc)x+bd.$$

The comparison of coefficients produce the following equations

$$a+c=-1, \quad ac+b+d=4, \quad ad+bc=3, \quad bd=5.$$

If try $b=1, d=5$, then the third equation yields $5a+c=3$. Considering the first equation, we obtain $a=1, c=-2$. By checking, the second equation is satisfied by the solution. So

$$x^4 - x^3 + 4x^2 + 3x + 5 = (x^2+x+1)(x^2-2x+5).$$

7. Given that y^2+3y+6 is a factor of the polynomial $y^4-6y^3+my^2+ny+36$. Find the values of constants m and n.

Suppose that $y^4 - 6y^3 + my^2 + ny + 36 = (y^2+3y+6)(y^2+ay+b)$, where a, b are constants to be determined below. From

$$(y^2+3y+6)(y^2+ay+b) = y^4+(a+3)y^3+(3a+b+6)y^2+(6a+3b)y+6b,$$

we have system of simultaneous equations

$$a+3=-6, \quad 3a+b+6=m, \quad 6a+3b=n, \quad 6b=36.$$

Therefore $a=-9$ and $b=6$. Then $m=-15, n=-36$.

Testing Questions (6-B)

1. Let $x^2+6x+1=u, x^2+1=v$, then

$$\begin{aligned}
&2(x^2+6x+1)^2+5(x^2+1)(x^2+6x+1)+2(x^2+1)^2\\
&=2u^2+5uv+2v^2=(2u+v)(u+2v)\\
&=[2(x^2+6x+1)+(x^2+1)][(x^2+6x+1)+2(x^2+1)]\\
&=(3x^2+12x+3)(3x^2+6x+3)=9(x^2+4x+1)(x+1)^2.
\end{aligned}$$

2. Let $x^4+ax^2+b=(x^2+2x+5)(x^2+cx+d)$. From

$$\begin{aligned}
&(x^2+2x+5)(x^2+cx+d)\\
&=x^4+(2+c)x^3+(2c+d+5)x^2+(5c+2d)x+5d,
\end{aligned}$$

we have the system of simultaneous equations

$$2+c=0,\quad 2c+d+5=a,\quad 5c+2d=0,\quad 5d=b,$$

hence $c=-2, d=5, a=6, b=25$. Thus, $a+b=31$.

3. $(ab+cd)(a^2-b^2+c^2-d^2)+(ac+bd)(a^2+b^2-c^2-d^2)$

$$\begin{aligned}
&=(ab+cd)[(a^2-d^2)-(b^2-c^2)]+(ac+bd)[(a^2-d^2)+(b^2-c^2)]\\
&=(ab+cd+ac+bd)(a^2-d^2)-(ab+cd-ac-bd)(b^2-c^2)\\
&=(a+d)(b+c)(a-d)(a+d)-(a-d)(b-c)(b-c)(b+c)\\
&=(a-d)(b+c)[(a+d)^2-(b-c)^2]\\
&=(a-d)(b+c)(a+b-c+d)(a-b+c+d).
\end{aligned}$$

4. We have

$$\begin{aligned}
(ay+bx)^3&=a^3y^3+3a^2bxy^2+3ab^2x^2y+b^3x^3,\\
(ax+by)^3&=a^3x^3+3a^2bx^2y+3ab^2xy^2+b^3y^3,\\
(a^3+b^3)(x^3+y^3)&=a^3x^3+a^3y^3+b^3x^3+b^3y^3,
\end{aligned}$$

therefore

$$\begin{aligned}
&(ay+bx)^3+(ax+by)^3-(a^3+b^3)(x^3+y^3)\\
&=3a^2bxy^2+3ab^2x^2y+3a^2bx^2y+3ab^2xy^2\\
&=3abxy(ay+bx+ax+by)=3abxy[(a(y+x)+b(x+y)]\\
&=3abxy(a+b)(x+y).
\end{aligned}$$

5. We have

$$\begin{aligned}
a^5b-ab^5&=ab(a^4-b^4)=ab(a-b)(a+b)(a^2+b^2),\\
b^5c-bc^5&=bc(b^4-c^4)=bc(b-c)(b+c)(b^2+c^2),\\
c^5a-ca^5&=ca(c^4-a^4)=ca(c-a)(c+a)(c^2+a^2).
\end{aligned}$$

If a, b have same parity, then $a - b$, $a + b$ and $a^2 + b^2$ are all even, so $a^5b - ab^5$ is divisible by 8.

If a and b have different parity, then the parity of c must be the same as a or b, say a and c have same parity, then $c^5a - ca^5$ is divisible by 8.

Solutions to Test Questions 7

Testing Questions (7-A)

1. For $x > 0$, $\dfrac{|x + |x||}{x} = \dfrac{|x + x|}{x} = \dfrac{2x}{x} = 2$;

   For $x < 0$, $\dfrac{|x + |x||}{x} = \dfrac{|x - x|}{x} = \dfrac{0}{x} = 0$.

2. $\dfrac{2x - 1}{3} - 1 \ge x - \dfrac{5 - 3x}{2} \Longleftrightarrow 2(2x - 1) - 6 \ge 6x - 3(5 - 3x)$

   $\Longleftrightarrow 7 \ge 11x \Longleftrightarrow x \le \dfrac{7}{11}$.

   For $x < -3$, $|x - 1| - |x + 3| = (1 - x) + (3 + x) = 4$, so its maximum value and minimum value are both 4.

   For $-3 \le x \le \dfrac{7}{11}$, $|x - 1| - |x + 3| = (1 - x) - (3 + x) = -2 - 2x$, so its maximum value is $-2 - 2(-3) = 4$, and its minimum value is $-2 - 2 \cdot \dfrac{7}{11} = -3\dfrac{3}{11}$.

   Thus, the maximum value is 4, and the minimum value is $-3\dfrac{3}{11}$.

3. For $x < 0$, $|1 - x| = 1 + |x| \Longleftrightarrow |x - 1| = 1 - x$, the answer is (D);

   For $0 \le x \le 1$, $|1 - x| = 1 + |x| \Longleftrightarrow 1 - x = 1 + x$, i.e. $x = 0$. The answer is (A), (B), (D);

   For $1 < x$, $|1 - x| = 1 + |x| \Longleftrightarrow x - 1 = 1 + x$, so no solution.

   Thus, the answer is (D).

4. For $x \le -1$, $|x + 1| + |x - 2| + |x - 3| = -(x + 1) - (x - 2) - (x - 3) = 4 - 3x \ge 7$;

   For $-1 < x \le 2$, $|x + 1| + |x - 2| + |x - 3| = (x + 1) - (x - 2) - (x - 3) = 6 - x \ge 4$, the minimum value is 4;

For $2 \le x < 3$, $|x+1|+|x-2|+|x-3| = (x+1)+(x-2)-(x-3) = x+2 \ge 4$, the minimum value is 4;

For $3 \le x$, $|x+1|+|x-2|+|x-3| = (x+1)+(x-2)+(x-3) = 3x-4 \ge 4$, the minimum value is 5.

Thus, the global minimum value is 4.

5. $\dfrac{||x|-2x|}{3} = \dfrac{|-x-2x|}{3} = \dfrac{-3x}{3} = -x.$

6. Let $S = |x-a|+|x-b|+|x-c|+|x-d|$. $|x-c|+|x-b|$ is minimum if and only if $b \le x \le c$, and similarly, $|x-a|+|x-d|$ is minimum if and only if $a \le x \le d$. Thus S is minimum if and only if $b \le x \le c$, and in the case

$$S = |c-b|+|d-a|.$$

Note: The same result can be obtained also if partition the number axis as five intervals : $x \le a$, $a \le x \le b$, $b < x < c$, $c \le x < d$, and $d \le x$, and then discuss the local minimum values on each interval.

7. When $a+b \ge 0$, we have $a+b = a-b$, so $b = 0$. When $a+b < 0$, then $-(a+b) = a-b$, so $a = 0$. Thus, $ab = 0$ in any case.

8. Only two cases are possible: $|a-b| = 1, |c-a| = 0$ or $|a-b| = 0, |c-a| = 1$, so it is always true that

$$|c-a|+|a-b|+|b-c| = 1+1 = 2.$$

9. $2a^3-3a^2-2a+1 = a(2a^2-3a-2)+1 = a(2a+1)(a-2)+1 > 0$ and $2a^3-3a^2-3a-2009 = a(a^2-3a-4) = a(a+1)(a-4) > 0$, so

$$\begin{aligned}&|2a^3-3a^2-2a+1|-|2a^3-3a^2-3a-2009|\\&=(2a^3-3a^2-2a+1)-(2a^3-3a^2-3a-2009)\\&=-2a+3a+2010 = a+2010 = 4019.\end{aligned}$$

10. $||x-a|-b| = 3 \Longleftrightarrow |x-a|-b = 3$ or $|x-a|-b = -3$.

If $|x-a| = 3+b$ has exactly one solution and $|x-a| = -3+b$ has two solutions, then $b = -3$ and $|x-a| = -6$ has two solutions, a contradiction.

Therefore $|x-a| = 3+b$ has two solutions and $|x-a| = -3+b$ has exactly one solution. Thus, $b = 3$ and the three distinct solutions are

$$x_1 = a, \quad x_2 = a+6, \quad x_3 = a-6.$$

Testing Questions (7-B)

1. We estimate the lower bound of n first. From $|x_1 + x_2 + \cdots + x_n| \geq 0$ and $|x_i| < 1$ for $i = 1, 2, \ldots, n$, we have

$$n > |x_1| + |x_2| + \cdots + |x_n| = 49 + |x_1 + x_2 + \cdots + x_n| \geq 49.$$

For $n = 50$, we take $x_1, x_2, \cdots, x_{50}$ by

$$x_i = \begin{cases} 0.98 & i = 1, 3, 5, \ldots, 49, \\ -0.98 & i = 2, 4, 6, \ldots, 50. \end{cases}$$

then $x_1, x_2, \ldots, x_{50}$ satisfy the requirement of the problem. Thus $n_{\min} = 50$.

2. Let $S = |x - a_1| + |x - a_2| + \cdots + |x - a_n|$. Since $|x - a| + |x - b|$ takes its minimum value $|b - a|$ if and only if x is between a and b.

If n is even, i.e. $n = 2k$ for some natural number k, we have

$$\begin{aligned} &|x - a_1| + |x - a_n| \geq a_n - a_1, \\ &|x - a_2| + |x - a_{n-1}| \geq a_{n-1} - a_2, \\ &\quad \vdots \\ &|x - a_{\frac{n}{2}}| + |x - a_{\frac{n}{2}+1}| \geq a_{\frac{n}{2}+1} - a_{\frac{n}{2}}. \end{aligned}$$

By adding them up, we obtain a lower bound

$$S \geq a_n + a_{n-1} + \cdots + a_{\frac{n}{2}+1} - a_{\frac{n}{2}} - a_{\frac{n}{2}-1} - \cdots - a_2 - a_1.$$

For any x satisfying $a_{\frac{n}{2}} \leq x \leq a_{\frac{n}{2}+1}$, above inequalities all become equalities, so the lower bound is the minimum value of S.

If n is odd, i.e. $n = 2k + 1$ for some natural number k, similar reasoning shows that when $x = a_{\frac{n+1}{2}} = a_{k+1}$, S takes its minimum value

$$a_n + a_{n-1} + \cdots + a_{\frac{n+1}{2}+1} - a_{\frac{n-1}{2}} - a_{\frac{n-3}{2}} - \cdots - a_2 - a_1.$$

3. Since in the simplified expression there is no x, it should be true that $|4 - 5x| = 4 - 5x$ and $|1 - 3x| = 3x - 1$, i.e.

$$1 - 3x \leq 0 \quad \text{and} \quad 4 - 5x \geq 0,$$

hence $\frac{1}{3} \leq x \leq \frac{4}{5}$. In fact, on this interval

$$2x + |4 - 5x| + |1 - 3x| + 4 = 2x + 4 - 5x + 3x - 1 + 4 = 7.$$

4. From $b + c = -a, a + c = -b, a + b = -c$ we have

$$x = -\left|\frac{|a|}{b+c} + \frac{|b|}{a+c} + \frac{|c|}{a+b}\right| = -\left|\frac{|a|}{-a} + \frac{|b|}{-b} + \frac{|c|}{-c}\right| = -1,$$

therefore

$$x^{2007} - 2007x + 2007 = -1 + 2007 + 2007 = 4013.$$

5. We suspect by considering 1 to 4, or 1 to 6 etc, that each pair (a_i, b_i) contains one number not less than 100 and one number not greater than 100. In fact, if $a_i > 100, b_i > 100$ for some natural number i, then the following $101(= 100 - (i - 1) + i = 101$ numbers

$$a_i, a_{i+1}, \dots, a_{100}, b_1, b_2, \dots, b_i$$

are all greater than 100, a contradiction. Similarly, there is no pair (a_i, b_i) with both a_i and b_i less than 100. Thus,

$$\begin{aligned}&|a_1 - b_1| + |a_2 - b_2| + \cdots + |a_{99} - b_{99}| + |a_{100} - b_{100}|\\&= 101 + 102 + \cdots + 200 - (1 + 2 + 3 + \cdots + 100)\\&= 100 \times 100 = 10000.\end{aligned}$$

Solutions to Testing Questions 8

Testing Question (8-A)

1. $|5x - 4| - 2x = 3$ yields $|5x - 4| = 2x + 3$, then $x \geq -\frac{3}{2}$ and

$$|5x - 4| = 2x + 3 \Longleftrightarrow 5x - 4 = 2x + 3 \text{ or } 5x - 4 = -2x - 3$$

$$\Longleftrightarrow x = \frac{7}{3} \text{ or } x = \frac{1}{7}.$$

2. When $a \leq -\frac{7}{2}$, then

$$|2a + 7| + |2a - 1| = 8 \Longleftrightarrow -(2a + 7) - (2a - 1) = 8 \Longleftrightarrow a = -\frac{7}{2},$$

no solution.

When $-\frac{7}{2} < a \leq \frac{1}{2}$, then

$|2a+7|+|2a-1|=8 \Longleftrightarrow (2a+7)-(2a-1)=8 \Longleftrightarrow 8=8$, so $a=-3,-2,-1,0$.

When $\frac{1}{2}<a$, then

$|2a+7|+|2a-1|=8 \Longleftrightarrow (2a+7)+(2a-1)=8 \Longleftrightarrow a=\frac{1}{2}$, no solution.

Thus, $a=-3,-2,-1,0$, the answer is (B).

3. $|x-|2x+1||=3 \Longleftrightarrow x-|2x+1|=3$ or $x-|2x+1|=-3$

$$\Longleftrightarrow 0 \le x-3=|2x+1| \text{ or } 0 \le x+3=|2x+1|.$$

From $x-3=|2x+1|$ we have $2x+1=x-3$ or $2x+1=3-x$, i.e. $x=-4$ or $x=\frac{2}{3}$, however, they are less than 3, so not applicable.
From $x+3=|2x+1|$ we have $2x+1=x+3$ or $2x+1=-x-3$, i.e. $x=2$ or $x=-\frac{4}{3}$.
Thus, the answer is (C).

4. Let x be a negative solution. Then $-x=ax+1$ i.e. $x=-\dfrac{1}{a+1}<0$. Therefore $a+1>0$, i.e. $a>-1$.

   On the other hand, if $x_1>0$ is a positive solution of the equation, then $x_1=ax_1+1 \Longleftrightarrow x_1=\frac{1}{1-a}>0 \Longleftrightarrow a<1$. Since the equation has no positive solution, we must have $a \ge 1$.

   Thus, $a \ge 1$, the answer is (C).

5. It is clear that $a \ge 0$, and then $||x-2|-1|=a \Longleftrightarrow |x-2|-1=a$ or $|x-2|-1=-a$.

   If $|x-2|-1=a$ has two integer solutions and $|x-2|-1=-a$ has only one integer solution, then $1-a=0$, i.e. $a=1$.

   If $|x-2|-1=a$ has only one integer solution and $|x-2|-1=-a$ has two integer solutions, then $1+a=0$, i.e. $a=-1$, a contradiction.

   Thus, $a=1$, the answer is (B).

6. For $x<0$, the equation becomes $-ax-2008x-2008^2=0$, i.e. $(a+2008)x=-2008^2$. Since the equation has negative solution, so $a+2008>0$, i.e. $a>-2008$.

The given equation cannot have solution $x = 0$. If the equation has a positive solution $x_0 > 0$, then

$ax_0 - 2008x_0 - 2008^2 = 0$, so $(a - 2008)x_0 = 2008^2$, then $a - 2008 > 0$, i.e. $a > 2008$. Thus, the range of a is $-2008 < a < 2008$.

7. (i) has no solution means $m > 0$; (ii) has exactly one solution means $n = 0$; and (iii) has two solutions means $k < 0$. Thus $m > n > k$, the answer is (A).

8. Let

$$|x - y| = x + y - 2, \tag{15.32}$$

$$|x + y| = x + 2. \tag{15.33}$$

From the equation (15.32) we find that $x + y - 2 \geq 0$, so $x + y \geq 2 > 0$. Then (15.33) becomes $x + y = x + 2$, i.e. $y = 2$. By substituting back $y = 2$ into the first equation, we have $|x - 2| = x$, i.e. $x - 2 = x$ or $x = 2 = -x$. The first one has no solution, and $x - 2 = -x$ has solution $x = 1$. Thus, the solution (x, y) for the original system is $(1, 2)$.

9. By factorization, from the given equation we have $(|x| + 3)(|x| - 2) = 0$. Since $|x| + 3 \geq 3$, we have $|x| = 2$. Therefore the roots are 2 and -2, their sum is 0, so the answer is (C).

10. From $2x + y = 6$ we have $y = 6 - 2x$. By substituting it into the first equation, we have

$x + 3(6 - 2x) + |3x - (6 - 2x)| = 19 \Longleftrightarrow |5x - 6| = 1 + 5x$

$\Longleftrightarrow 5x - 6 = -5x - 1$ or $5x - 6 = 1 + 5x$ (no solution)

$\Longleftrightarrow x = \dfrac{5}{10} = \dfrac{1}{2}, y = 6 - 2x = 5.$

Thus, the solution is $x = \dfrac{1}{2}, y = 5$.

Testing Questions (8-B)

1. $|x - 2y| = 1 \Longleftrightarrow x = 1 + 2y$ or $x = 2y - 1$. By substituting it into the second equation, we have

$$|1 + 2y| + |y| = 2$$

and

$$|2y - 1| + |y| = 2.$$

From $|1 + 2y| + |y| = 2$ we obtain $y = -1$ or $y = \frac{1}{3}$. Then correspondingly, $x = 1 + 2y = -1$ or $x = \frac{5}{3}$.

From $|2y - 1| + |y| = 2$ we obtain $y = 1$ or $y = -\frac{1}{3}$. Then correspondingly, $x = 2y - 1 = 1$ or $x = -\frac{5}{3}$. Thus, the solutions are

$$\begin{cases} x = -1, \\ y = -1. \end{cases} \quad \begin{cases} x = \frac{5}{3}, \\ y = \frac{1}{3}. \end{cases} \quad \begin{cases} x = 1, \\ y = 1. \end{cases} \quad \begin{cases} x = -\frac{5}{3}, \\ y = -\frac{1}{3}. \end{cases}$$

2. Since $||||x - 1| - 1| - 1| - 1| = 0 \iff |||x - 1| - 1| - 1| = 1$ and $|||x - 1| - 1| - 1| = 1 \iff ||x - 1| - 1| = 2$ or $||x - 1| - 1| = 0$ $\iff |x - 1| - 1 = \pm 2$ or $|x - 1| - 1 = 0 \iff |x - 1| = 3$ or -1 or 1 $\iff x = 0, \pm 2, 4$.

   Thus, the answer is (A).

3. We remove the outer absolute value signs by taking squares to both sides of the equation, then

$$(|a| - (a + b))^2 < (a - |a + b|)^2,$$
$$a^2 - 2|a|(a + b) + (a + b)^2 < a^2 - 2a|a + b| + (a + b)^2,$$
$$|a|(a + b) > a|a + b| \iff a \neq 0,$$

   so $a + b > \frac{a}{|a|} \cdot |a + b|$. Hence $a < 0, a + b > 0$. Thus, $b > -a > 0$, the answer is (B).

4. (i) When $a = \frac{1}{26}$, the equation becomes an identity for all $x \neq 2$. When $a \neq \frac{1}{26}$, the equation is equivalent to

$$x - 2 = x - 52a \quad \text{or} \quad x - 2 = -(x - 52a).$$

   The first equation has no solution. The second equation has the solution $x = \frac{52a + 2}{2} = 26a + 1$.

(ii) When $a = p^2$, where p is an odd prime number, then, from the result of (i),

$$x = 26a + 1 = 26p^2 + 1.$$

If $p = 3$, then $x = 26 \cdot 9 + 1 = 235$ which is composite.

If $p = 3k \pm 1$, then $x = 26(3k \pm 1)^2 + 1 = 234k^2 \pm 156k + 1 = 3(78k^2 \pm 52k + 9)$ which is composite also.

5. If $a_1 > a_2 > a_3 > a_4$, then we can remove all the absolute value signs, such that the system becomes as follows:

$$(a_1 - a_2)x_2 + (a_1 - a_3)x_3 + (a_1 - a_4)x_4 = 1 \tag{15.34}$$
$$(a_1 - a_2)x_1 + (a_2 - a_3)x_3 + (a_2 - a_4)x_4 = 1 \tag{15.35}$$
$$(a_1 - a_3)x_1 + (a_2 - a_3)x_2 + (a_3 - a_4)x_4 = 1 \tag{15.36}$$
$$(a_1 - a_4)x_1 + (a_2 - a_4)x_2 + (a_3 - a_4)x_3 = 1. \tag{15.37}$$

(15.35) – (15.34) yields

$$(a_1 - a_2)x_1 - (a_1 - a_2)x_2 - (a_1 - a_2)x_3 - (a_1 - a_2)x_4 = 0,$$

so

$$x_1 - x_2 - x_3 - x_4 = 0. \tag{15.38}$$

Similarly, (15.36) – (15.35) and (15.37) – (15.36), after simplification, yield

$$x_1 + x_2 - x_3 - x_4 = 0 \tag{15.39}$$

and

$$x_1 + x_2 + x_3 - x_4 = 0, \tag{15.40}$$

respectively. Then (15.39) – (15.38) yields $x_2 = 0$, and (15.40) – (15.39) yields $x_3 = 0$. Thus, from (15.40), $x_1 = x_4$.

Finally, from (15.37), we obtain $x_1 = x_4 = \dfrac{1}{a_1 - a_4}$, $x_2 = x_3 = 0$.

If $a_2 > a_1$, we can exchange the subscripts 1 and 2 of a_1 and a_2, and exchange the subscripts 1 and 2 of x_1 and x_2. The system does not change. Continuing this process if necessary, until we have $a_1 > a_2 > a_3 > a_4$.

Thus, among a_1, a_2, a_3, a_4, if a_i is maximum and a_j is minimum, then

$$x_i = x_j = \frac{1}{a_i - a_j}, \quad \text{and the other two are zeros.}$$

Solutions to Testing Questions 9

Testing Questions (9-A)

1. From $(n-2)\times 180° < 2007°$ we have $n-2 < 12$, i.e. $n < 14$.

   When $n = 13$, the sum of interior angles of convex 13-sided polygon is $11 \times 180° = 1980°$, so the maximum value of n is 13.

2. Let $\angle B = x$, then $\angle AQP = 2x = \angle QAP$, so $\angle QPA = 180 - 4x$.

   Further,

$$\because \angle APC = \angle ACP = 3x,$$
$$\therefore 2 \times 3x + x = 180°,$$
$$\therefore x = \frac{180}{7} = 25\frac{5}{7}.$$

   Thus, the answer is (A).

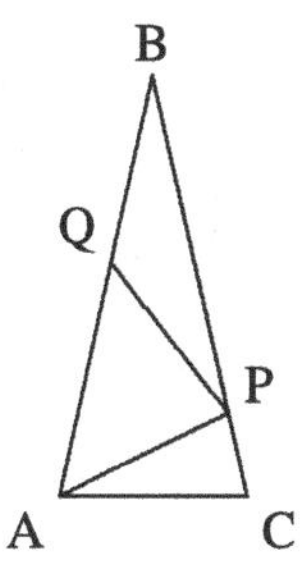

3. From $AE = AC$ and $BC = BF$, we have

$$\angle AEC = \frac{1}{2}(180° - \angle A) = 90° - \frac{1}{2}\angle A,$$
$$\angle BFC = \frac{1}{2}(180° - \angle B) = 90° - \frac{1}{2}\angle B,$$

   Therefore

$$\angle ECF = 180° - \angle AEC - \angle BFC = \frac{1}{2}(\angle A + \angle B) = 45°.$$

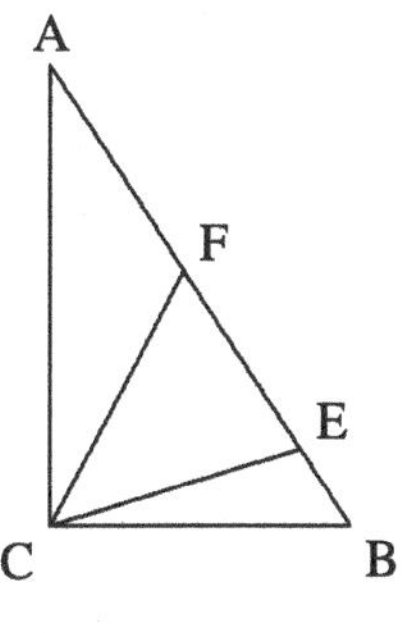

4. Let the lengths of the three sides be a, b, c respectively, where $a \geq b \geq c$. $c \leq \frac{17}{3} < 6$ leads $a - b < c \leq 5$. We classify the triangles according to the integral value of c for counting.

   (i) When $c = 1$, then $a + b = 16$, $a - b = 0$, therefore $a = b = 8$, $c = 1$ is a solution;

   (ii) When $c = 2$, then $a + b = 15$, $a - b = 1$, therefore $a = 8$, $b = 7$, $c = 1$ is a solution;

(iii) When $c = 3$, then $a+b = 14,\ a-b = 0$, or $a+b = 14,\ a-b = 2$, therefore $a = b = 7,\ c = 3$ and $a = 8, b = 6, c = 3$ are 2 solutions;

(iv) When $c = 4$, then $a+b = 13,\ a-b = 1$, or $a+b = 13,\ a-b = 3$, therefore $a = 7, b = 6, c = 4$ and $a = 8, b = 5, c = 4$ are 2 solutions;

(v) When $c = 5$, then $a+b = 12,\ a-b = 0$, or $a+b = 12,\ a-b = 2$, therefore $a = b = 6,\ c = 5$ and $a = 7, b = c = 5$ are 2 solutions;

Thus, there are 8 such triangles in total.

5. When $b = 2$ then $a = 1$. From $a > c - b$ we have $c = b$. Since $b < c$, we have no required solution. Thus, the answer is 0.

6. Let the lengths of two legs of the right angle be a and b where $a = 21$, and let c be the length of the hypotenuse. Then $c^2 - b^2 = 21^2$, i.e. $(c-b)(c+b) = 3^2 \cdot 7^2$. To let the sum $21 + b + c$ be minimum, $b + c$ should be minimum, therefore $c - b$ should be maximum. Thus, $c - b = 9,\ c + b = 49$, i.e. the perimeter is $21 + 49 = 70$.

7.

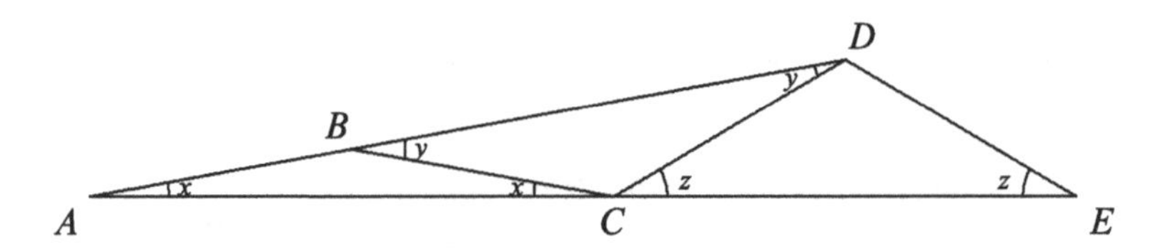

Let $x = \angle BAC = \angle BCA, y = \angle CBD = \angle CDB, z = \angle DCE = \angle CED$, then

$$y = 2x, \quad z = x + y = 3x.$$
$$\because x + 140^\circ + z = 180^\circ \Longrightarrow 140^\circ + 4x = 180^\circ,$$
$$\therefore x = 10^\circ, \quad \text{i.e. } \angle EAD = 10^\circ.$$

8. From $\angle A = 80^\circ$ and $AB = AC$, we have $\angle B = \angle C = 50^\circ$. Then

$$\angle CDE = \frac{1}{2}(180^\circ - 50^\circ) = 65^\circ,$$
$$\angle FDB = \frac{1}{2}(180^\circ - 50^\circ) = 65^\circ,$$
$$\therefore \angle EDF = 180^\circ - 2 \times 65^\circ = 50^\circ.$$

The answer is (C).

9. $n + (n + 1 + (n + 2) \le 100$ yields $1 \le n \le 32$. The triangle inequality means $n + 2 < 2n + 1$ which implies $n \ge 2$. The triangle is acute yields $n^2 + (n + 1)^2 > (n + 2)^2$,

$$2n^2 + 2n + 1 > n^2 + 4n + 4,$$
$$n^2 - 2n - 3 > 0,$$
$$(n - 3)(n + 1) > 0, \quad \therefore n > 3.$$

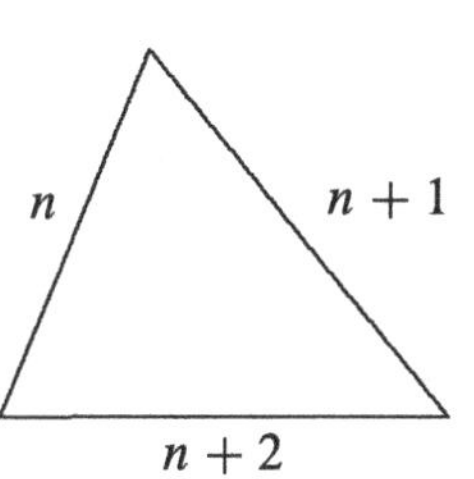

Thus, there are in total 29 such triangles.

10. In $\triangle ABC$ and $\triangle ABD$, since $AB = AC = BD$, we have

$$\angle C = \frac{1}{2}(180° - \angle BAC),$$
$$\angle D = \tfrac{1}{2}(180° - \angle DBA),$$
$$\therefore \angle C + \angle D = 180° - \frac{1}{2}(\angle BAC + \angle DBA).$$
$$\because \angle BAC + \angle DBA = 90°,$$
$$\therefore \angle C + \angle D = 180° - 45° = 135°,$$

the answer is (D).

Testing Questions (9-B)

1. Connect CH. As shown in the digram, let the areas of triangles be $S_0, S_1, \cdots, S_4$. Without loss of genrality we may assume that $S_0 = 1$. Since $AH/HE = 3$ yields $S_1 = 3$, then $\dfrac{BH}{HD} = \dfrac{5}{3}$ implies that $S_2 = \dfrac{3}{5}S_1 = \dfrac{9}{5}$.

Since $\dfrac{S_2 + S_3}{S_4} = \dfrac{3}{1}$, so $S_4 = \dfrac{1}{3}(S_2 + S_3)$.

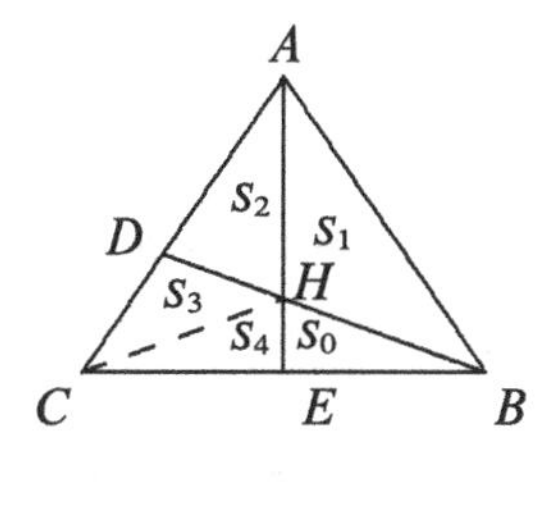

$$\therefore \frac{S_3}{S_4 + S_0} = \frac{S_3}{\frac{1}{3}(S_2 + S_3) + S_0} = \frac{3}{5},$$
$$\frac{S_3}{\frac{1}{3}(\frac{9}{5} + S_3) + 1} = \frac{3}{5},$$

Hence, $S_3 = \frac{6}{5}$, $S_4 = \frac{1}{3}\left(\frac{9}{5} + \frac{6}{5}\right) = 1$, so $S_0+S_1 = 4$, $S_2+S_3+S_4 = 4$, i.e. $CE = BE$, the triangle ABC is isosceles. Thus,

$$\angle C = \frac{1}{2}(180^\circ - \angle A) = \frac{1}{2}(180^\circ - 70^\circ) = 55^\circ.$$

The answer is (B).

2. Since A_1B and A_1C bisect $\angle ABC$ and $\angle ACD$ respectively, $\angle A = \angle ACD - \angle ABC = 2(\angle A_1CD - \angle A_1BC) = 2\angle A_1$, therefore $\angle A_1 = \frac{1}{2}\angle A$.

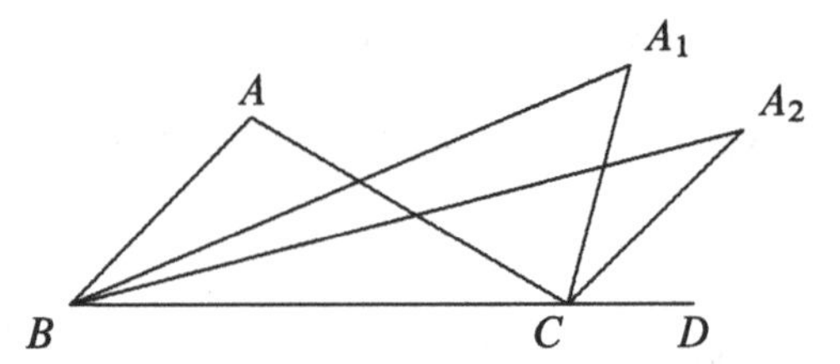

Similarly, we have $A_{k+1} = \frac{1}{2}A_k$ for $k = 1, 2, 3, 4$. Hence

$$A_5 = \frac{1}{2}A_4 = \frac{1}{4}A_3 = \frac{1}{2^3}A_2 = \frac{1}{2^4}A_1 = \frac{1}{2^5}A = \frac{96^\circ}{32} = 3^\circ.$$

3. As shown in the digram, we have $\angle B = \angle C$.

Further, we have

$$\begin{aligned}&\angle DEB - \angle CFE\\ &= \angle FEB - \angle CFE - 60^\circ = \angle C - 60^\circ,\\ &\angle ADF - \angle DEB\\ &= \angle ADE - \angle DEB - 60^\circ = \angle B - 60^\circ,\\ &\therefore \angle DEB - \angle CFE = \angle ADF - \angle DEB,\end{aligned}$$

i.e. $\angle DEB = \angle\frac{1}{2}(\angle ADF + \angle CFE)$.

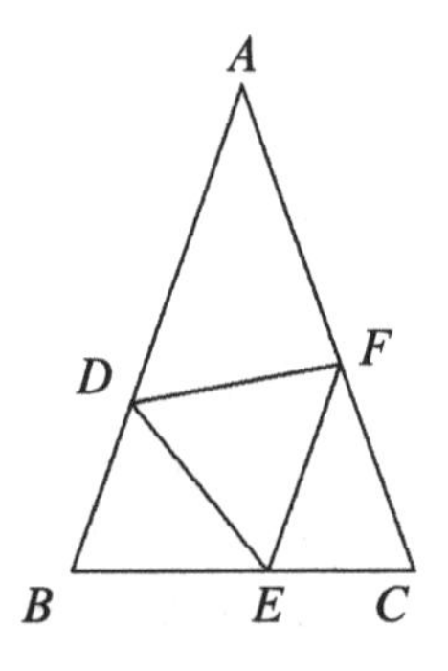

4. Let $a = DE$, $b = BE = AC$. $\because \frac{AC}{BC} = \frac{DE}{BE}$, $\therefore \frac{b}{1-a} = \frac{a}{b}$ and

$$b^2 = (1-a)a = a - a^2, \text{i.e.} a^2 + b^2 = a.$$

$$\because\ a^2 + b^2 = (\frac{1}{2})^2,$$

$$\therefore\ a = \frac{1}{4} = \frac{1}{2}BD, \text{ thus } \angle B = 30^\circ.$$

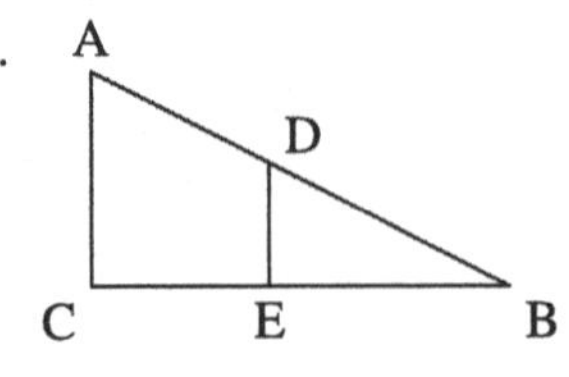

5. From $\angle BNA = 180° - 50° - 80° = 50° = \angle BAN$,

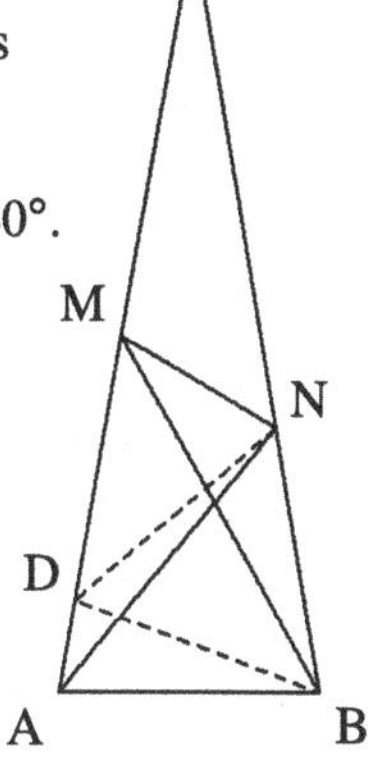

we have $AB = NB$. On AC take D such that $\angle ABD = 20°$, then $\angle ADB = \angle DAB = 80°$, therefore $DB = AB = NB$. Since $\angle DBN = 60°$, the triangle NDB is equilateral, therefore

$$ND = DB = NB, \quad \angle MDN = 180° - 80° - 60° = 40°.$$
$$\because \angle DBM = 60° - 20° = 40° \text{ and}$$
$$\angle DMB = 180° - 60° - 80° = 40°,$$

it follows that $DM = DB = DN$, therefore

$$\angle DMN = \angle DNM = 70°, \text{ so}$$
$$\angle NMB = \angle DMN - \angle DMB = 70° - 40° = 30°.$$

Solutions to Testing Questions 10

Testing Question (10-A)

1. From A introduce $AE \perp BC$ at E. Since $\angle B = \angle C = 45°$, $\angle BAE = \angle CAE = 45°$, $\therefore AE = BE = CE$. By Pythagoras' Theorem,

$$\begin{aligned} & BD^2 + CD^2 \\ &= (BE + ED)^2 + (CE - ED)^2 \\ &= BE^2 + 2BE \cdot DE + DE^2 \\ &\quad + CE^2 - 2CE \cdot DE + DE^2 \\ &= BE^2 + CE^2 + 2DE^2 \\ &= 2(AE^2 + DE^2) = 2AD^2. \end{aligned}$$

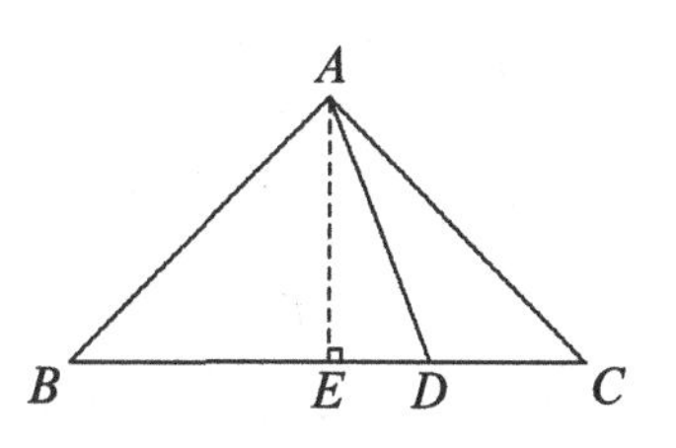

2. Suppose that $\angle C = 90°$. Let $a = BC$, $b = AC$, $c = AB$. Then

$$a + b + \sqrt{a^2 + b^2} = 30, \tag{15.41}$$
$$\frac{ab}{2} = 30. \tag{15.42}$$

From (15.42), $ab = 60$, therefore, from (15.41),

$$\sqrt{(a+b)^2 - 120} = 30 - (a+b),$$
$$(a+b)^2 - 120 = 900 + (a+b)^2 - 60(a+b),$$
$$\therefore\ (a+b) = 17.$$

By substituting $b = 17 - a$ into $ab = 60$, it follows that $a^2 - 17a + 60 = 0$, so $a = 5, b = 12$ or $a = 12, b = 5$. By Pythagoras' Theorem, $c = \sqrt{5^2 + 12^2} = \sqrt{169} = 13$, i.e. the lengths of three side are $5, 12, 13$ respectively.

3. From D introduce $DE \perp AB$ at E. By symmetry we have $DE = DC$ and $AE = AC = 9$ cm, and hence $EB = 6$ cm. Let $CD = 3x$ cm, then $BD = 5x$ cm, therefore

$$(5x)^2 - (3x)^2 = 6^2,$$
$$(4x)^2 = 6^2,$$
$$\therefore\ x = \frac{6}{4} = \frac{3}{2}.$$

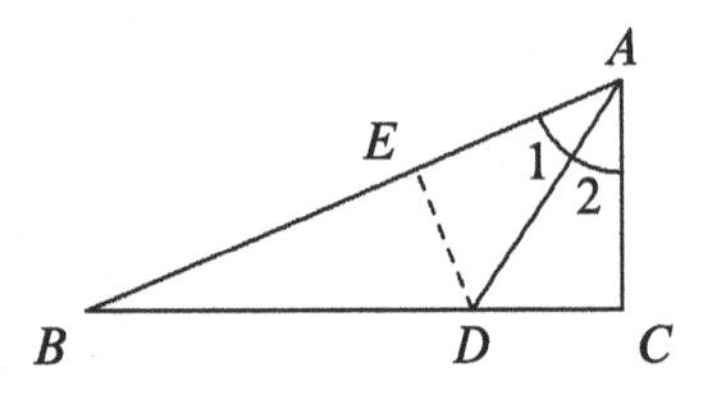

Thus, $DE = CD = \dfrac{9}{2}$ cm.

4. Connect AE. Let $CE = x$ cm. From $AE = EB = 12 - x$ cm,

Since $AE^2 = AC^2 + CE^2$,

$$(12 - x)^2 = 6^2 + x^2,$$
$$144 + x^2 - 24x = 36 + x^2,$$
$$24x = 108,$$
$$\therefore\ x = 4.5.$$

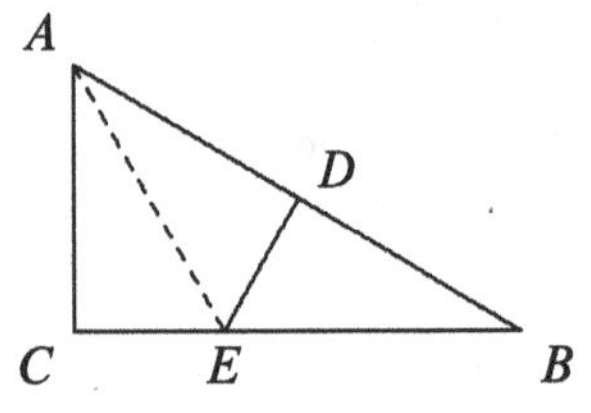

Thus, $CE = 4.5$ cm.

5. Suppose that BD and AC intersect at O. From $BE = \frac{1}{4}BD$ we have $OE = \frac{1}{2}BD - \frac{1}{4}BD = \frac{1}{4}BD$. Let $OE = x$, then $OC = OD = 2x$.

From $OC^2 - OE^2 = CE^2$,

$$(2x)^2 - x^2 = 5^2,$$
$$x^2 = \frac{25}{3},$$
$$\therefore\ AC = 4x = \frac{20}{\sqrt{3}} = \frac{20\sqrt{3}}{3}\ \text{cm}.$$

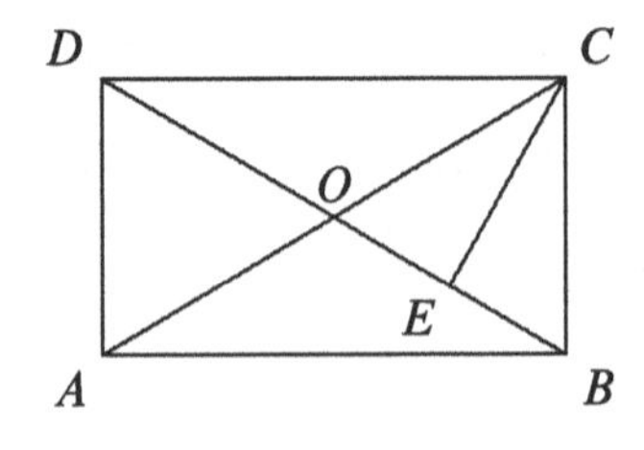

6. From $BD^2 = BC^2 + DC^2 = BC^2 + \frac{1}{4}AC^2$, we have

$$\begin{aligned} 4BD^2 &= 4BC^2 + AC^2 \\ &= 4BC^2 + AB^2 - BC^2 \\ &= AB^2 + 3BC^2. \end{aligned}$$

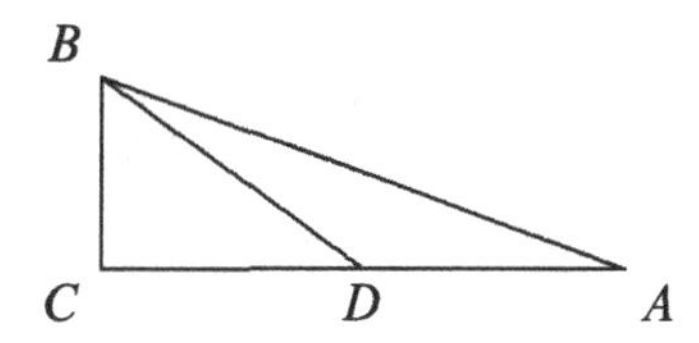

7. By using the Pythagoras' Theorem,

$$\begin{aligned} &AD^2 + BE^2 \\ &= AC^2 + CD^2 + CE^2 + BC^2 \\ &= (AC^2 + BC^2) + (CD^2 + CE^2) \\ &= AB^2 + DE^2. \end{aligned}$$

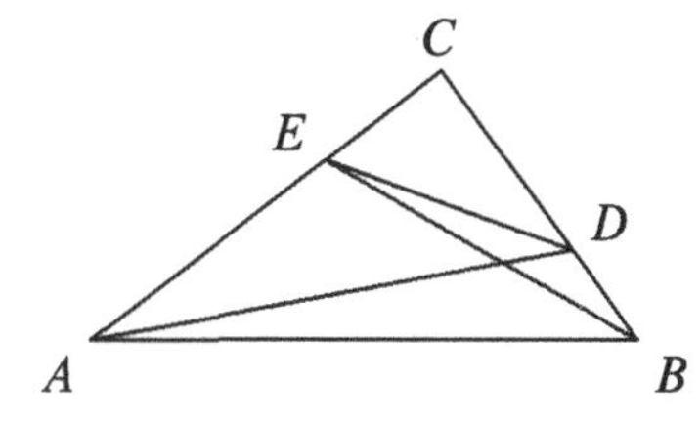

8. From A introduce $AD \perp BC$ at D. Then $BD = DC$. Let $BD = DC = x$ and $DP_i = x_i$,
   By Pythagoras' Theorem, for $1 \le i \le 100$,

$$\begin{aligned} m_i &= AP_i^2 + BP_i \cdot P_iC \\ &= AP_i^2 + (x - x_i)(x + x_i) \\ &= AP_i^2 - x_i^2 + x^2 \\ &= AD^2 + x^2 = AB^2 = 4. \end{aligned}$$

   Thus,

$$m_1 + m_2 + \cdots + m_{100} = 400.$$

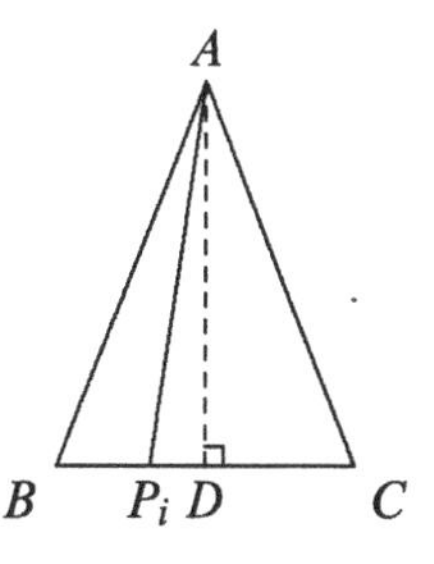

9. From A introduce $AG \parallel CB$, intersecting the extension of FD at G, connect EG.
   By symmetry, we have $DG = DF, AG = BF$, so ED is the perpendicular bisector of FG. Thus,

$$\begin{aligned} &EF^2 = EG^2 \\ &= AE^2 + AG^2 \\ &= AE^2 + BF^2. \end{aligned}$$

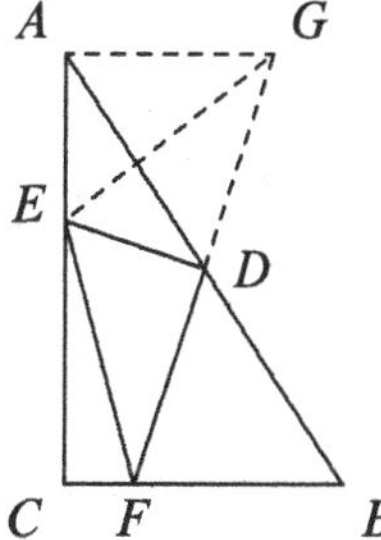

10. Rotate $\triangle BPA$ around B in anti-clockwise direction by 60°, then $A \to C$.

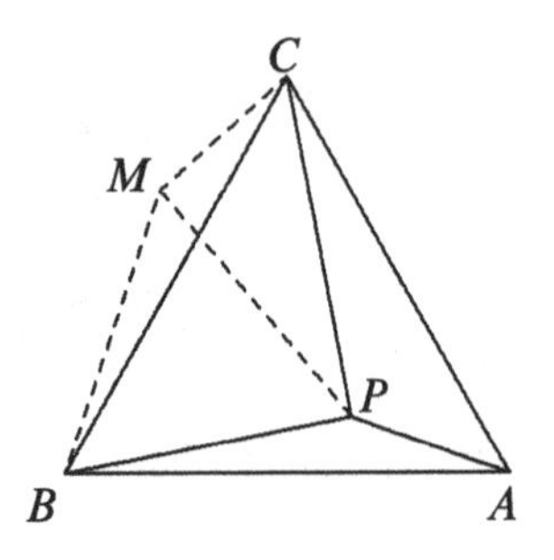

Let the image of P be M under the rotation. Then $BM = BP, \angle MBP = 60°$, so $\triangle MBP$ is equilateral, i.e. $MP = 2\sqrt{3}$. From $MC = PA = 2$ and

$$MP^2 + MC^2 = 12 + 4 = 4^2 = PC^2,$$

so $\angle PMC = 90°, \angle BPA = \angle BMC = 150°$. Further, $PC = 2MC$ implies $\angle MPC = 30°$, so $\angle BPC = 90°$, and

$$BC^2 = PB^2 + PC^2 = 12 + 16 = 28, \quad \text{thus, } BC = \sqrt{28} = 2\sqrt{7}.$$

Testing Questions (10-B)

1. Suppose that $OD \perp AB$ at D, then

$$AD = BD = \frac{1}{2}(63 + 33) = 48 \text{ (cm)},$$

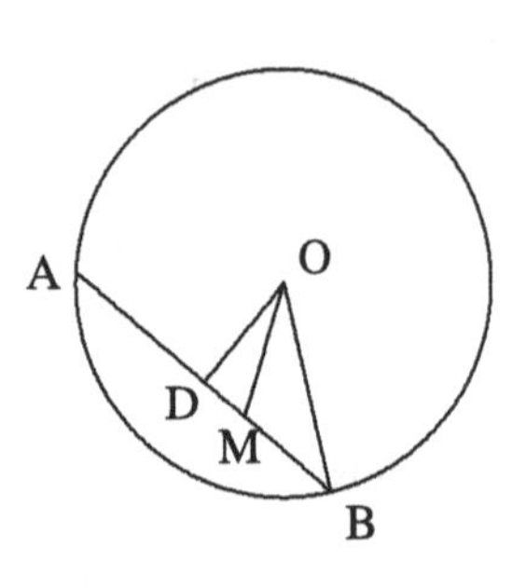

therefore $MD = 48 - 33 = 15$ (cm), hence

$$\begin{aligned} OD^2 &= OB^2 - BD^2 \\ &= 52^2 - 48^2 = 400 \text{ (cm}^2\text{)}, \\ \therefore OM &= \sqrt{OD^2 + MD^2} \\ &= \sqrt{625} = 25 \text{ (cm)}. \end{aligned}$$

2. By passing through P introduce the lines $QR \parallel BC$, where Q and R are on AB and DC respectively. Then

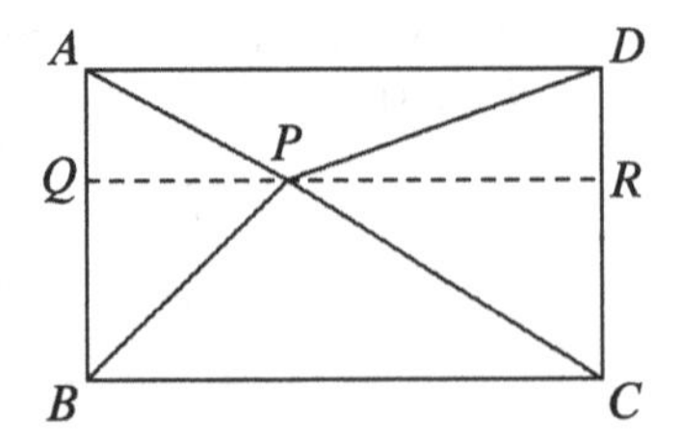

$$\begin{aligned} &AP^2 + PC^2 \\ &= PQ^2 + AQ^2 + PR^2 + CR^2 \\ &= PQ^2 + RD^2 + PR^2 + BQ^2 \\ &= (PQ^2 + BQ^2) + (PR^2 + RD^2) \\ &= PB^2 + PD^2. \end{aligned}$$

Thus, $PD^2 = PA^2 + PC^2 - PB^2$
$= 9 + 25 - 16 = 18$, i.e. $PD = \sqrt{18}$.

3. Suppose that such a triangle exists. Let a and b be the lengths of the two legs of the right angle, where $b = ka$ for some positive integer k. Let c be the length of the hypotenuse. By Pythagoras' Theorem,

$$c^2 = a^2 + b^2 = (1 + k^2)a^2.$$

Since $\left(\frac{c}{a}\right)^2 = 1+k^2$ is an integer, so $a \mid c$. Let $\frac{c}{a} = m$, then $m^2 = 1+k^2$. However, $k^2 < 1 + k^2 < (k + 1)^2$ indicates that $1 + k^2$ is not a perfect square, a contradiction.

Thus, there is no a triangle satisfying the required conditions.

4. The given conditions implies that $\angle BCE = 30°$, so $CE = 2BE = 12$. By Pythagoras' Theorem, $CB = \sqrt{CE^2 - BE^2} = \sqrt{3 \cdot 6^2} = 6\sqrt{3}$. Similarly, $CD = \sqrt{3} \cdot DF = \sqrt{3}(6\sqrt{3} - 2) = 18 - 2\sqrt{3}$.

Therefore the area of $ABCD$ is $[ABCD] = 6\sqrt{3}(18 - 2\sqrt{3}) = 108\sqrt{3} - 36 < 216 - 36 = 180$. From

$$150 < 108\sqrt{3} - 36 \iff 186^2 < 3(108)^2 \iff 34596 < 34992,$$

$\therefore [ABCD] > 150$. The answer is (E).

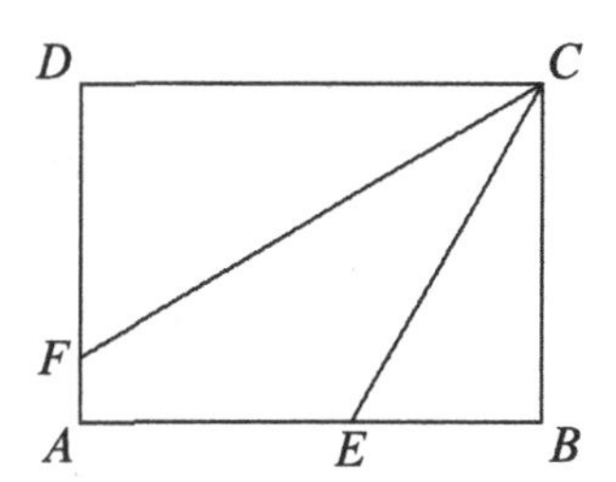

5. **Necessity** If $AC \perp BD$, O is the point of intersection of AC and BD, then

$$\begin{aligned} &AB^2 + DC^2 \\ &= (AO^2 + BO^2) + (CO^2 + DO^2) \\ &= (AO^2 + DO^2) + (BO^2 + CO^2) \\ &= AD^2 + BC^2. \end{aligned}$$

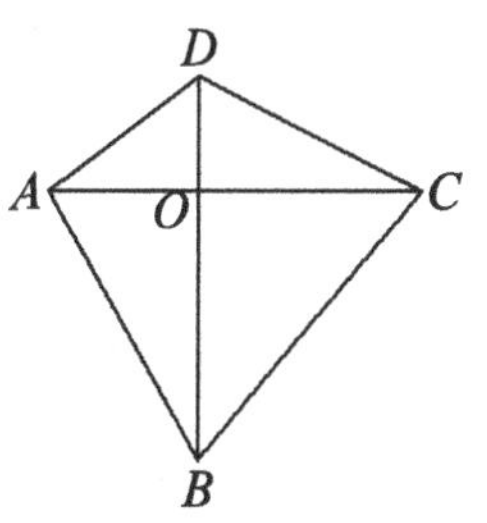

Sufficiency If $AB^2+CD^2 = AD^2+BC^2$, then $AB^2 - AD^2 = BC^2 - DC^2$. If A', C' be the perpendicular projections of A, C on BD respectively, then

$$\begin{aligned} &AB^2 - AD^2 = A'B^2 - A'D^2 = BD \cdot (A'B - A'D), \\ &BC^2 - DC^2 = BC'^2 - C'D^2 = BD \cdot (BC' - C'D), \\ &\therefore BA' - A'D = BC' - C'D, \text{i.e. } A' \text{ coincides with } C'. \end{aligned}$$

Thus, $AC \perp BD$.

Solutions to Testing Questions 11

Testing Questions (11-A)

1. $\because \angle ABC = 180° - 75° - 60° = 45°$, we have $\angle BAD = 45°$, therefore $AD = BD$. Since

$$\angle HBD = 90° - \angle ACB = \angle CAD,$$

we have $\triangle HBD \cong \triangle CAD$ (A.S.), hence

$$HD = CD, \quad \therefore \angle CHD = 45°.$$

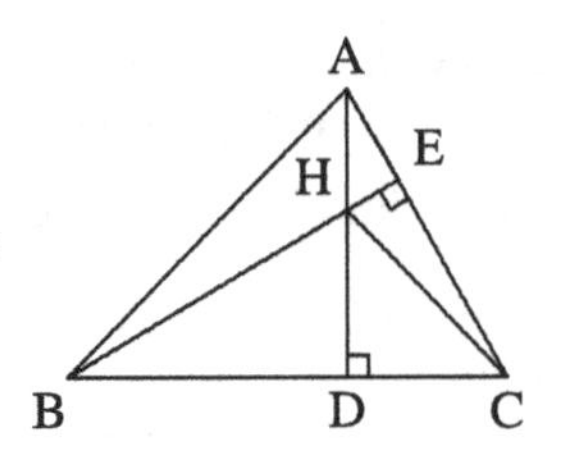

2. Connect CD and connect PD.

$\because AD = BD,\ CD = CD,\ CA = CB,$
$\therefore \triangle CDA \cong \triangle CDB,$ (S.S.S.),
$\therefore \angle DCB = \angle DCA = 30°.$
$\because BP = BC,$ and
$\angle DBC = \angle DBP,$
$\therefore \triangle DBP \cong \triangle DBC$ (S.A.S.),
$\therefore \angle BPD = \angle BCD = 30°.$

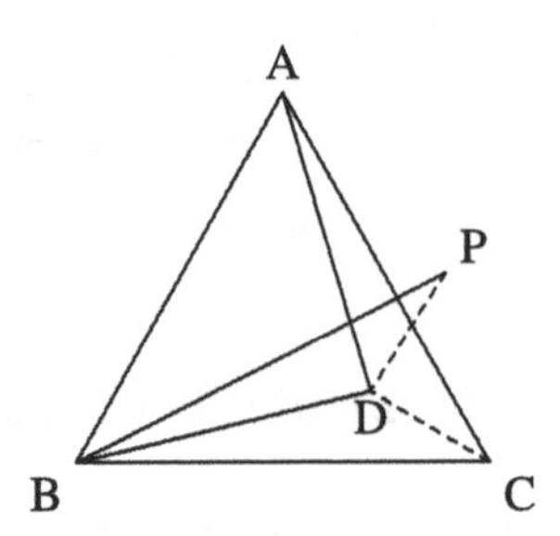

3. Let $a = AP,\ b = AQ$. Extend AD to P' such that $DP' = PB$, then $\text{Rt}\triangle PBC \cong \text{Rt}\triangle P'DC$ (S.S.), so $CP' = PC$ and

$$\begin{aligned} QP' &= (1-b) + (1-a) \\ &= 2 - (a+b) = PQ, \end{aligned}$$

$\therefore \triangle CQP \cong \triangle CQP'$ (S.S.S.),
$\therefore \angle PCQ = \angle P'CQ.$
$\because \angle PCB = \angle P'CD,$
$\therefore \angle PCP' = \angle DCB = 90°,$
$\therefore \angle PCQ = \frac{1}{2} \cdot 90° = 45°.$

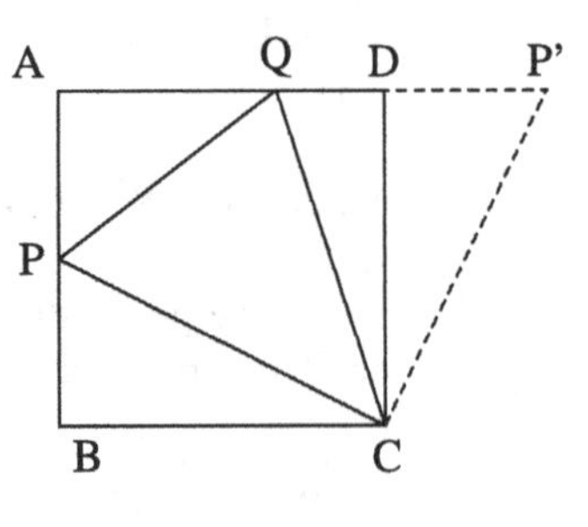

4. $BE = CF$ and $BC = CD$ leads to $\text{Rt}\triangle DFC \cong \text{Rt}\triangle CEB$ (S.S.), therefore $\angle CDF = \angle BCE$ which implies $CE \perp DF$.

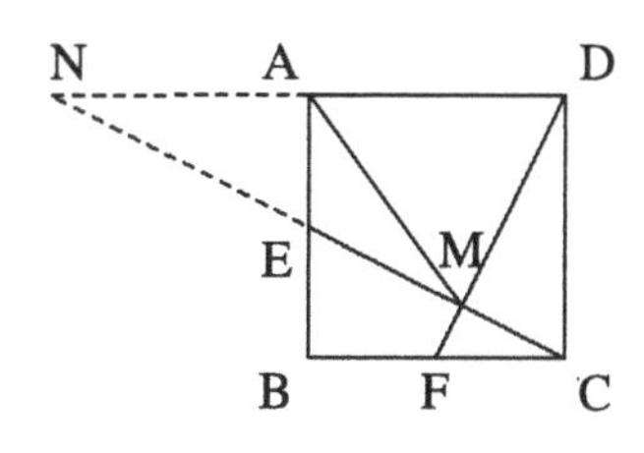

Extend CE to intersect the extension of DA at N.
$\because AE = BE$ and $\angle NEA = \angle BEC$,
$\therefore \text{Rt}\triangle AEN \cong \text{Rt}\triangle BEC$ (S.A.),
therefore $AN = BC = AD$, i.e. AM is the median on the hypotenuse ND of the $\text{Rt}\triangle NMD$, hence $AM = AD = AN$.

5. Suppose that the lines AE and BC intersect at F. Then $DE = CE$ and

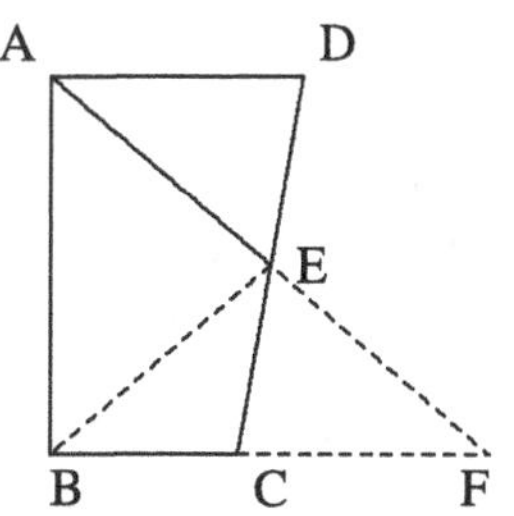

$$\angle ADE = \angle FCE, \angle AED = \angle CEF,$$
$$\therefore \triangle AED \cong \triangle FEC \quad \text{(A.A.S.)},$$

hence $AE = EF$.
Thus, BE is the median on the hypotenuse AF of the $\text{Rt}\triangle ABF$, we have

$$BE = EF = AE,$$
$$\therefore \angle DAE = \angle EFB = \angle EBF = \angle BEC.$$

Thus, $\angle AEB = \angle EBF + \angle EFB = 2\angle DAE$ implies that

$$\angle AEC = \angle AEB + \angle BEC = 3\angle DAE, \text{ i.e. } \angle DAE = \frac{1}{3}\angle AEC.$$

6. Introduce $MP \parallel AC$ and connect AP. Suppose that CM, AP intersect at O, then both triangles OMP and OAC are equilateral. We have

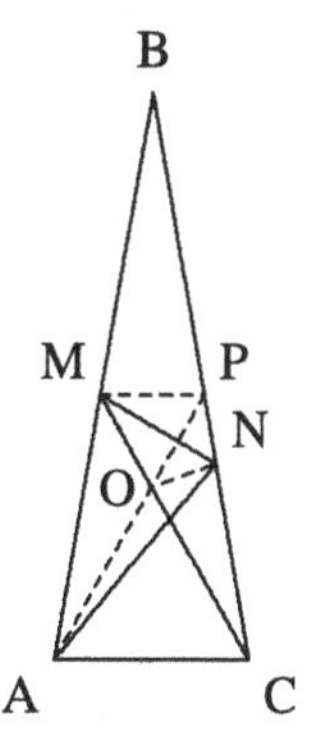

$$\angle ANC = 180^\circ - 80^\circ - 50^\circ = 50^\circ,$$
$$\therefore OC = AC = CN.$$
$$\because \angle NCO = 80^\circ - 60^\circ = 20^\circ,$$
$$\angle NOC = \tfrac{1}{2}(180^\circ - 20^\circ) = 80^\circ,$$
$$\therefore \angle PON = 120^\circ - 80^\circ = 40^\circ.$$
$$\because \angle OPN = 180^\circ - 60^\circ - 80^\circ = 40^\circ,$$

we obtain $ON = PN$, hence $\triangle ONM \cong \triangle PNM$ (S.S.S.), i.e.

$$\angle NMC = \angle NMO = \frac{1}{2}\angle PMO = 30^\circ.$$

Note: This question is the same as the Testing Question 5 of (9-B). Here we prefer the readers solve it again by using the congruence of triangles.

7. Suppose that the lines AE, BC intersect at F. From $AC = BC$ and that

$$\angle FAC = 90° - \angle AFC = \angle DBC,$$

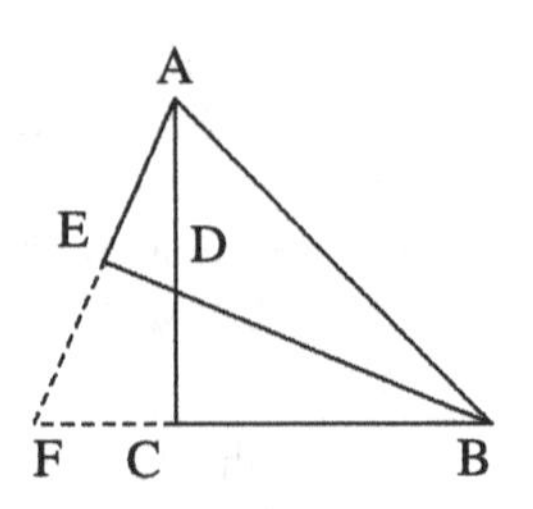

we have Rt$\triangle FAC \cong$ Rt$\triangle DBC$ (S.A.), therefore

$$AF = BD = 2AE, \text{ i.e. } AE = EF.$$

Thus, Rt$\triangle AEB \cong$ Rt$\triangle FEB$ (S.S.), hence

$$\angle ABE = \angle FBE.$$

8. Let E be the midpoint of BC. Connect EA, EP, and introduce $EF \perp AP$ at F.

By symmetry,

$$\triangle ABE \cong \triangle ADQ,$$
$$\therefore \angle BAE = \angle DAQ = \alpha, \text{so} \angle PAE = \alpha.$$
$$\therefore \triangle ABE \cong \triangle AFE \text{ (A.S.)}.$$

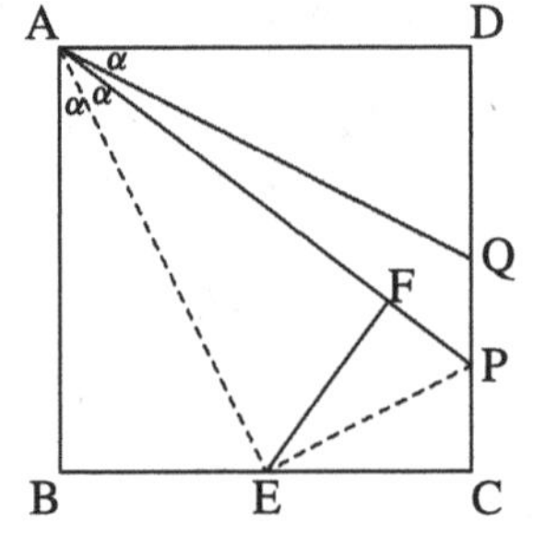

For Rt$\triangle EFP$ and Rt$\triangle ECP$, we have

$$EF = BE = EC \text{ and } EP = EP,$$
$$\therefore \triangle EFP \cong \triangle ECP \text{ (S.S.)},$$
$$\therefore PC = FP = 10 - 8 = 2.$$

9. Connect AC, AD and extend CB to P such that $BP = DE$. Connect AP.

In right triangles APB and ADE,

$$AB = AE, \text{ and } BP = DE$$
$$\Longrightarrow \triangle APB \cong \triangle ADE, \text{ (S.S.)}$$
$$\Longrightarrow AP = AD, \ CP = BC + DE = CD$$
$$\Longrightarrow \triangle ACD \cong \triangle ACP \text{ (S.S.S.)}.$$

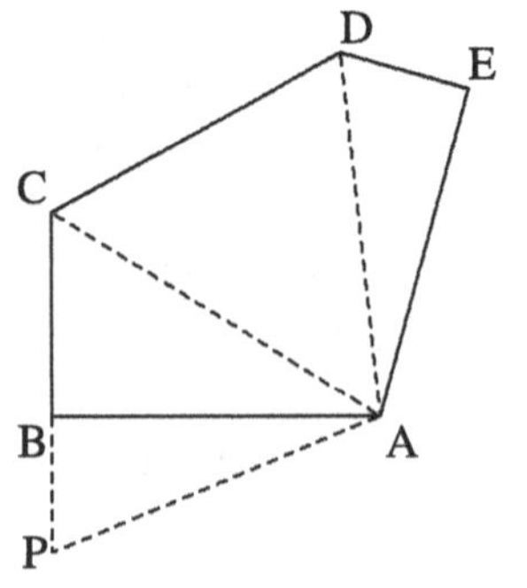

Therefore the height of $\triangle ACD$ is equal to AB, i.e. 1,

$$[ABCDE] = 2[ACD] = 2 \cdot \frac{1}{2} \cdot 1 \cdot 1 = 1.$$

10. When rotating the triangle ADC around A by 60° in anti-clockwise direction, the triangle $AD'B$ is obtained, as shown in the digram below.

From $\triangle AD'B \cong \triangle ADC$ we have

$$D'B = DC, \quad AD' = AD, \quad \angle DAD' = 60°,$$

so $\triangle AD'D$ is equilateral. Hence $D'D = AD$.
Thus, $\angle DD'B = 150° - 60° = 90°$, i.e. $\triangle BD'D$ is the right triangle, and is formed by the segments AD, BD, CD.

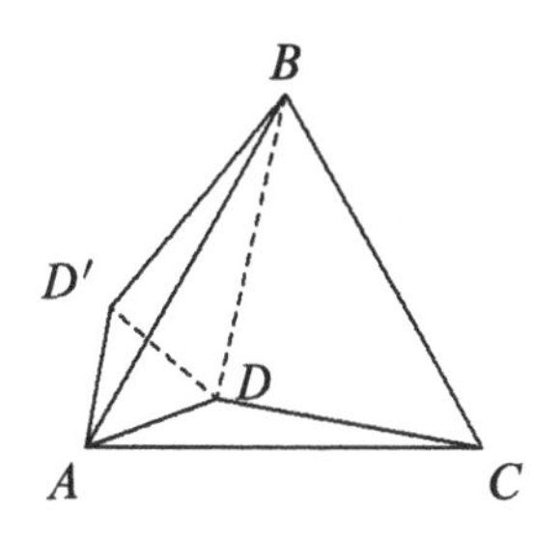

Testing Questions (11-B)

1. From the points A, C, E respectively introduce perpendiculars AA_1, CC_1, EE_1 to ℓ, where A_1, C_1, E_1 are on ℓ.

It is easy to see that

$\triangle AA_1B \cong \triangle BC_1C$,
$\triangle CC_1D \cong \triangle DE_1E$.
$\therefore A_1B = CC_1 = DE_1$.

Since the projection on ℓ of M is the midpoint of A_1E_1, so is also the midpoint of BD.

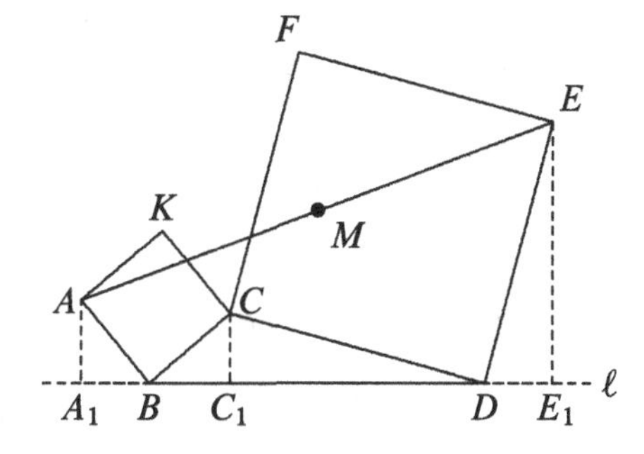

On the other hand, the distance from M to ℓ, is the midline of the trapezium AA_1E_1E, so it is

$$\frac{1}{2}(AA_1 + EE_1) = \frac{1}{2}(BC_1 + C_1D) = \frac{BD}{2}.$$

Thus, the point M is fixed even if C is changing.

2. From F introduce $FH \perp AB$ at H. Then $\angle ACF = \angle AHF = 90°$, hence $\triangle ACF \cong \triangle AHF$ (S.A.).
$\therefore CF = FH$.
$\because \angle ACD = 90° - \angle A = \angle B$,
$\therefore \angle FEC = \angle ACD + \frac{1}{2}\angle A$
$= \angle B + \frac{1}{2}\angle A = \angle CFE$,
$\therefore CE = CF = FH. \because CE \parallel FH$,
$\therefore \triangle ECG \cong \triangle HFB$ (S.A.).

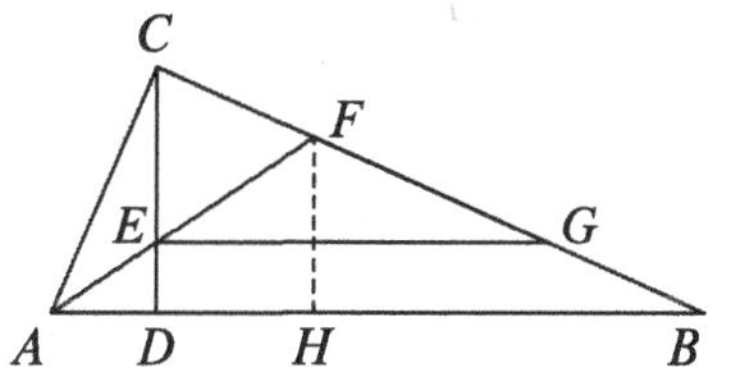

Thus, $CG = FB$, so that $CF = CG - FG = FB - FG = GB$.

3. Let AD be the angle bisector of $\angle BAC$, where AD intersects BC at D.

From D introduce $DE \perp AC$, intersecting AC at E. Then

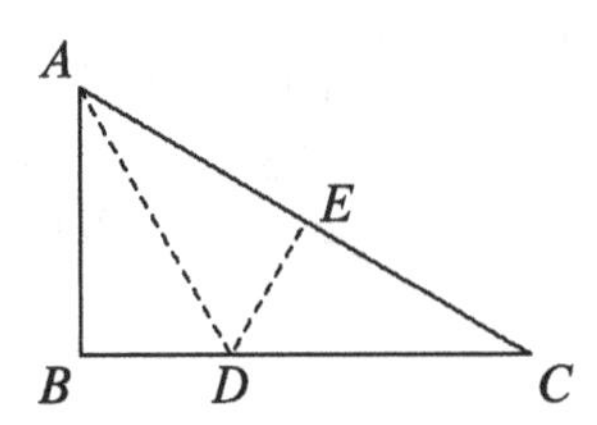

$$\begin{aligned}
&\angle DAC = \tfrac{1}{2}\angle A = \angle C,\\
&\therefore \triangle DEA \cong \triangle DEC,\\
&\Longrightarrow AE = EC = AB\\
&\Longrightarrow \triangle DAE \cong \triangle DAB \text{ (S.A.S.)}\\
&\Longrightarrow \angle ABD = \angle AED = 90^\circ.
\end{aligned}$$

Thus, $AB \perp BC$ at B.

4. Connect BD and extend BC to E such that $CE = CD$. Connect DE. From given conditions triangles ABD and CDE are both equilateral.

Since $\angle ADB = \angle CDE = 60^\circ$, we have

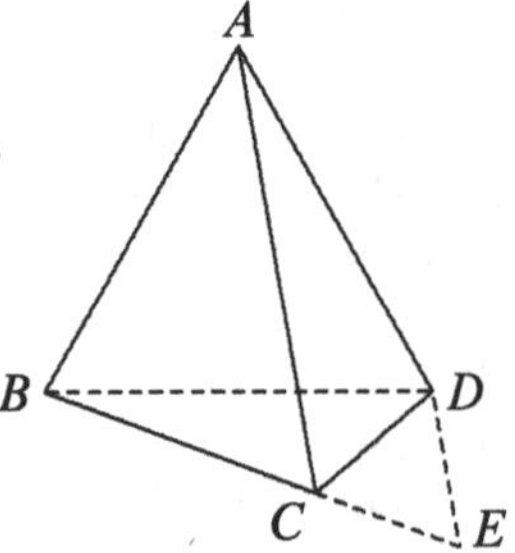

$$\begin{aligned}
&\angle ADC = 60^\circ + \angle BDC = \angle BDE.\\
&\because BD = AD, CD = ED,\\
&\therefore \triangle ADC \cong \triangle BDE \text{ (S.A.S.)},\\
&\therefore AC = BE = BC + CE\\
&\qquad = BC + CD.
\end{aligned}$$

5. Inside the region of $\angle ABC$ introduce the segment BQ, such that $\angle CBQ = \angle BAC = 20^\circ$ and $AB = BQ$. Connect AQ, CQ.

Since $\angle ABQ = 80^\circ - 20^\circ = 60^\circ$, $\triangle ABQ$ is equilateral, so $AQ = AB = AC$.

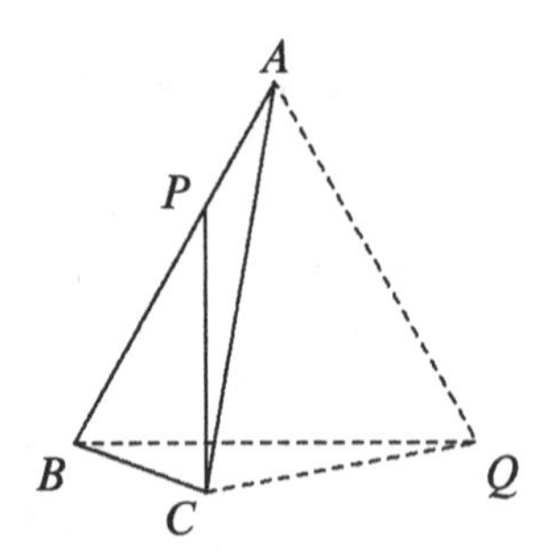

$$\begin{aligned}
&\because \angle CAQ = 60^\circ - 20^\circ = 40^\circ,\\
&\angle AQC = \tfrac{1}{2}(180^\circ - 40^\circ) = 70^\circ,\\
&\therefore \angle BQC = 70^\circ - 60^\circ = 10^\circ.\\
&\because \angle ACP = 30^\circ - 20^\circ = 10^\circ = \angle BQC,\\
&\triangle ACP \cong \triangle BQC \text{ (A.S.A.)},\\
&\therefore AP = BC.
\end{aligned}$$

Solutions to Testing Questions 12

Testing Questions (12-A)

1. Let P be the midpoint of the diagonal BD. Connect PE, PF. Then, by the midpoint theorem,

$$PE = \frac{1}{2}AB, PF = \frac{1}{2}CD.$$

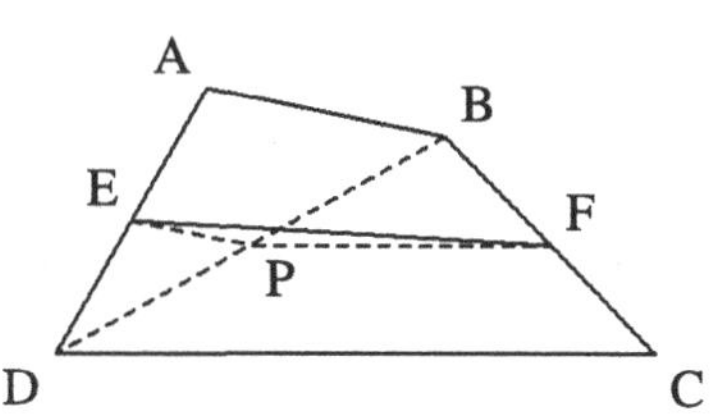

Applying triangle inequality to $\triangle PEF$, we have

$$EF < \frac{1}{2}(AB + CD).$$

2. Let F be the midpoint of AB and E the point of intersection of the lines BC and AD, as shown in the diagram on the right. From the midpoint theorem, $DC \parallel AB$ and $DC = \frac{1}{2}AB$ implies D, C are mid points of EA and EB respectively, so by this theorem again, D, M, F are collinear and F, N, C are collinear, and

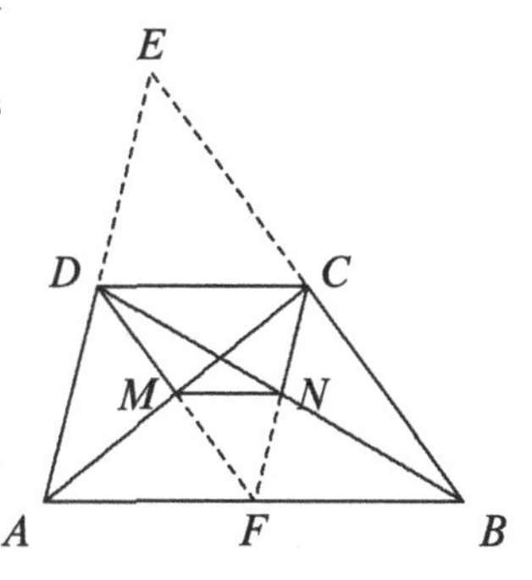

$$DM = \tfrac{1}{2}EC = \tfrac{1}{2}BC = MF,$$
$$FN = \tfrac{1}{2}AD = \tfrac{1}{2}DE = NC.$$
$$\therefore MN = \tfrac{1}{2}AF = \tfrac{1}{2}CD.$$

$\therefore l_1 = AB + BC + CD + AD = 2(CD + DM + MN + NC) = 2l_2$, i.e. $n = 2$.

3. From C introduce $CG \parallel BD$ such that CG

intersects the line AF produced at G. Then

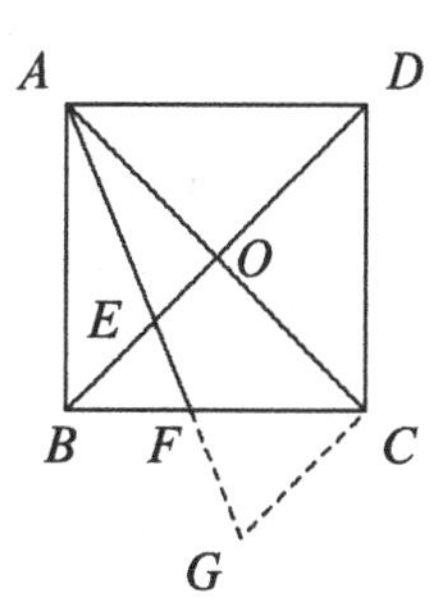

$$\begin{aligned} CG &= 2OE. \\ \because \angle CFG &= \angle AFB = 90° - 22.5° \\ &= 67.5°, \\ \angle CGF &= 180° - 45° - 67.5° \\ &= 67.5°, \\ \therefore CF &= CG = 2OE. \end{aligned}$$

4. Let F be the midpoint of AC. Connect DF, EF. From midpoint theorem,

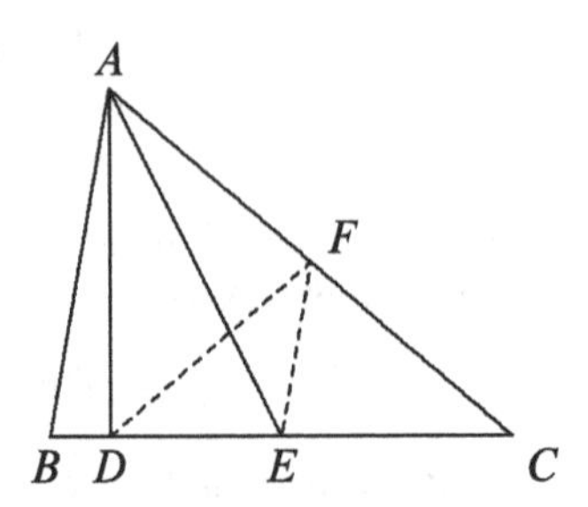

$$EF = \frac{1}{2}AB = ED,$$
$$\therefore \angle DFE = \angle EDF.$$
$$\because AF = FC \text{ and } \angle ADC = 90°,$$
$$\therefore DF = AF = FC,$$
$$\therefore \angle C = \angle EDF = \frac{1}{2}\angle CEF = \frac{1}{2}\angle B,$$
$$\therefore \angle B = 2\angle C.$$

5. It suffices to show $PQ = PR = QR$. Connect BP, CQ. Since $\triangle ABO$ and $\triangle CDO$ are both equilateral,

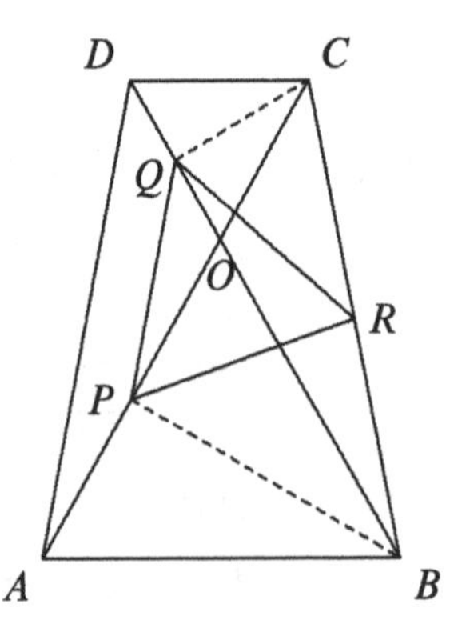

$$BP \perp AO, \quad CQ \perp DO.$$

Therefore PR and QR are both medians on hypotenuse of right triangles,

$$PR = BR = CR = QR.$$

By using the midpoint theorem,

$$PQ = \frac{1}{2}AD = \frac{1}{2}BC = PR = QR.$$

Thus, $\triangle PQR$ is equilateral.

6. From D introduce $DF \parallel BE$, intersecting AC at F. Then

$$EF = FC, \quad DF = \frac{1}{2}BE = 2.$$

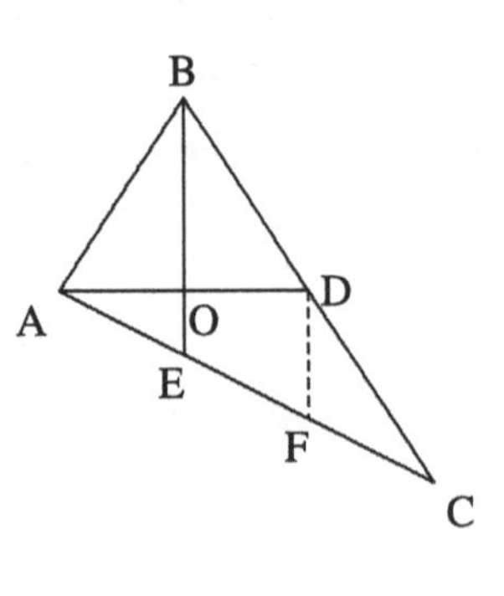

Since $\angle ABO = \angle DBO$ and $BO = BO$,

$$\text{Rt}\triangle ABO \cong \text{Rt}\triangle DBO \text{ (S.A.)},$$
$$\therefore AO = OD = 2, \quad OE = \tfrac{1}{2}DF = 1,$$

hence $BO = 3$.
By the Midpoint Theorem and the Pythagoras' Theorem
$FC = EF = AE = \sqrt{AO^2 + OE^2} = \sqrt{5}$,

$$AC = 3AE = 3\sqrt{5}, \quad AB = \sqrt{BO^2 + AO^2} = \sqrt{13},$$
$$BC = 2BD = 2AB = 2\sqrt{13}.$$

7. Extend BD, CE to P, Q respectively such that $BD = PD$ and $CE = QE$. Connect CP, AP, BQ, AQ. Then

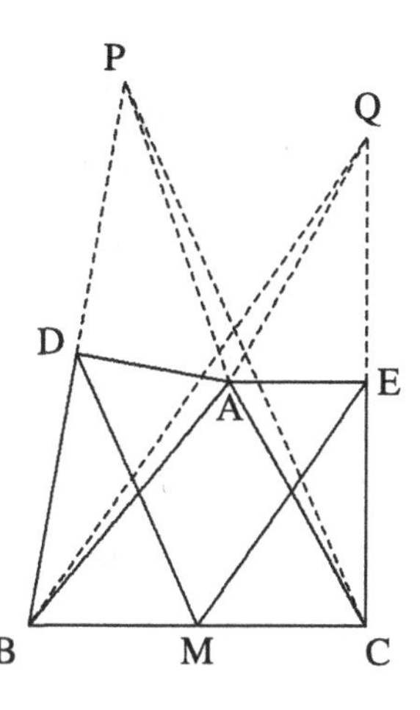

$$PC = 2DM, \qquad BQ = 2EM,$$

it suffices to show $PC = BQ$. Since AD, AE are the perpendicular bisectors of BP, CQ respectively,

$$AP = AB \text{ and } AC = AQ.$$

Further,

$$\begin{aligned}\angle PAC &= 360^\circ - 2\angle DAB - \angle BAC \\ &= 360^\circ - 2\angle EAC - \angle BAC \\ &= \angle BAQ,\end{aligned}$$

$\triangle PAC \cong \triangle BAQ$ (S.A.S.), $\therefore PC = BQ$.

Testing Questions (12-B)

1. The two angles $\angle AHE$ and $\angle BGE$ are on different sides of the line EF, it is needed to collect them together for their comparison. Here the midpoint theorem plays important role.

Connect AC, let P be the midpoint of AC. Connect PE, PF. By the midpoint theorem,

$$PE \parallel BG, PF \parallel AH,$$
$$\therefore \angle PFE = \angle AHE, \angle PEF = \angle BGE.$$

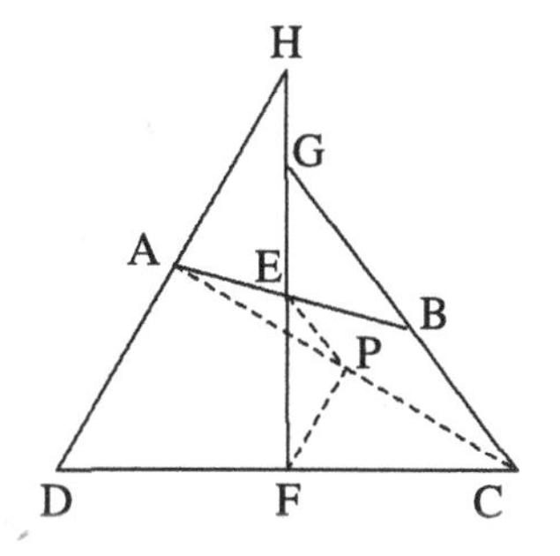

Thus, the $\angle AHE$ and $\angle BGE$ are replaced by the $\angle PFE$ and $\angle PEF$ respectively.

$$\because PE = \tfrac{1}{2}BC < \tfrac{1}{2}AD = PF,$$
$$\therefore \angle PFE < \angle PEF.$$
$$\therefore \angle AHE < \angle BGE.$$

2. Let G be the midpoint of BC, then

$$GN = BN - BG = \frac{1}{2}(BM - BC) = \frac{1}{2}CM,$$

therefore it suffices to show that $GN = CF$. Connect EG, CD and let K be the point of intersection of CD and EG. Connect KH.

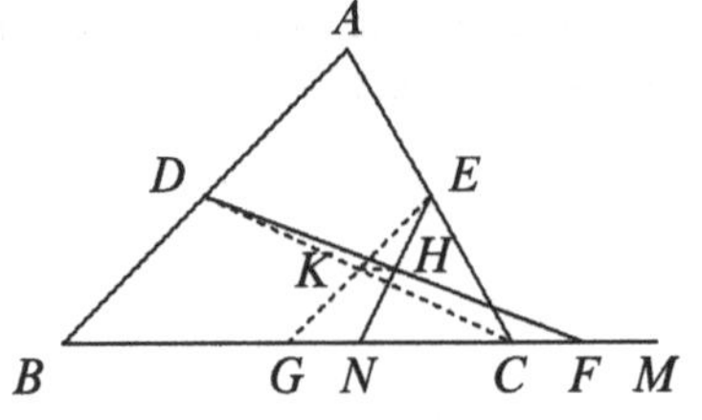

$$\because EK = KG \quad \text{and} \quad EH = HN,$$
$$\therefore KH \parallel BM, KH = \frac{1}{2}GN,$$
$$\therefore DH = HF. \text{ Then}$$
$$KH = \frac{1}{2}CF, \therefore GN = CF.$$

3. Connect CD. By symmetry,

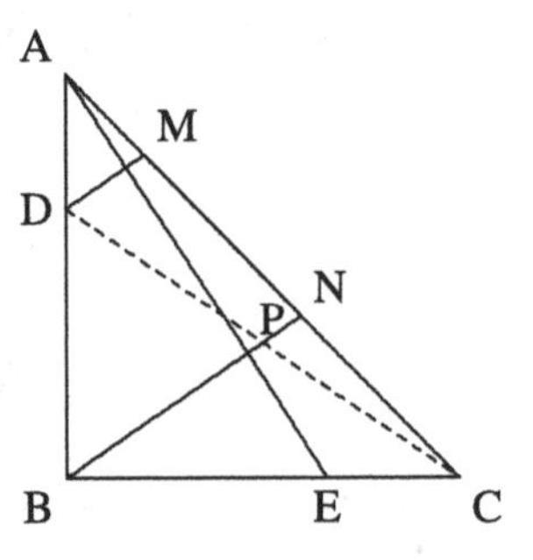

$$\triangle ABE \cong \triangle CBD \text{ (S.S.)},$$
$$\therefore \angle BAE = \angle BCD.$$
$$\because BN \perp AC, \ \therefore \angle BAE = \angle CBN,$$
$$\therefore \angle CBN = \angle BCD, \ \angle NBD = \angle CDB.$$

Let P be the point of intersection of BN and CD, then P is the midpoint of CD. Since $DM \parallel PN$, by the midpoint theorem,

$$MN = NC.$$

4. Let F be the midpoint of AB. Connect EF.

$$\because AD \parallel BC, \angle B + \angle A = 180^\circ,$$
$$\therefore \angle AEB = 180^\circ - 90^\circ = 90^\circ,$$
$$\therefore EF = \tfrac{1}{2}AB = BF = AF,$$
$$\therefore \angle AEF = \angle EAF = \angle EAD,$$
$$\text{i.e. } EF \parallel AD,$$
$$\therefore E \text{ is the midpoint of } CD.$$
$$\because EF \text{ is the middle line of the trapezium,}$$

therefore

$$\frac{1}{2}(AD + BC) = EF = \frac{1}{2}AB,$$

thus, $AB = AD + BC$.

Solutions to Testing Questions 13

Testing Questions (13-A)

1. (C). Since $\triangle BEF \sim \triangle BDC$ and $\triangle CEF \sim \triangle CAB$,

$$\frac{EF}{DC} = \frac{BF}{BC}, \quad \frac{EF}{AB} = \frac{CF}{BC},$$

$$\therefore EF\left(\frac{1}{80} + \frac{1}{20}\right) = 1,$$

$$\frac{EF}{16} = 1, \quad \therefore EF = 16.$$

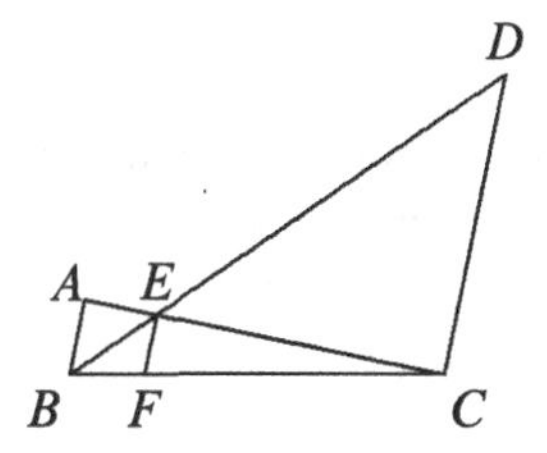

2. (C). Let $PC = x$. From $\triangle PAB \sim \triangle PCA$, we have

$$\frac{PA}{PB} = \frac{PC}{PA} = \frac{CA}{AB}$$

$$= \frac{6}{8} = \frac{3}{4}.$$

$$\therefore PC = \frac{9}{16}PB = \frac{9(x+7)}{16},$$

$$16x = 9x + 63$$

$\therefore x = 9$, the answer is (C).

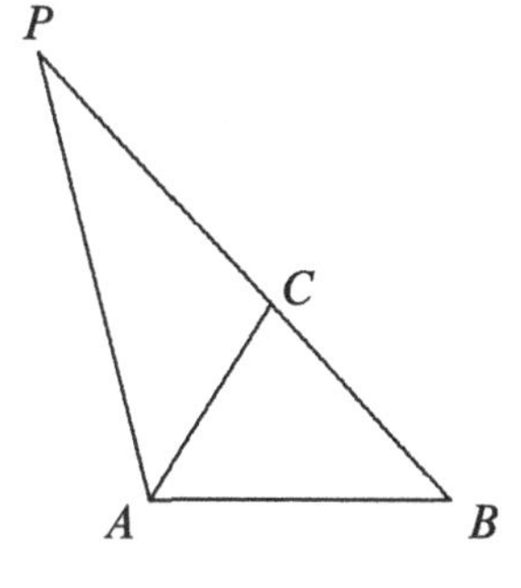

3. (A). From $\angle BAD = \angle EBF$ we have Rt$\triangle ABD \sim$ Rt$\triangle EBF$. Then

$$2 = \frac{AB}{BD} = \frac{BF}{EF}.$$

$$\because \triangle EFC \sim \triangle ABC,$$

$$\therefore EF = FC, \frac{BF}{FC} = 2,$$

$$\therefore \frac{EF}{AB} = \frac{FC}{BC} = \frac{1}{3},$$

$$\therefore EF = \frac{1}{3}AB = \frac{1}{3}a.$$

The answer is (A).

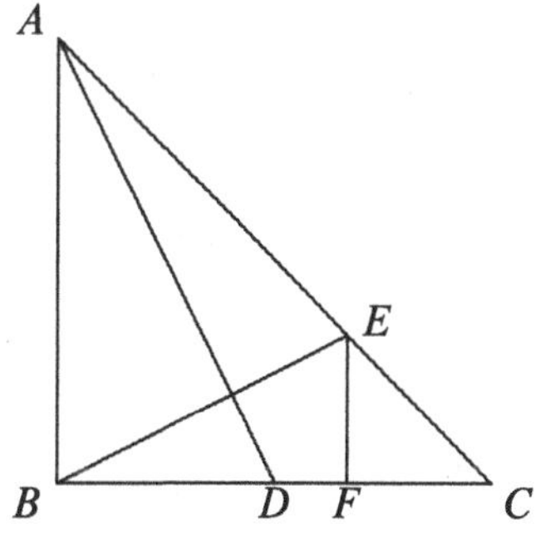

4. (B). $\because \angle MCN = \angle A = \angle B = 45°, \triangle MCN \sim \triangle CAN \sim \triangle MBC$,

$\therefore \dfrac{BC}{x+n} = \dfrac{x+m}{AC}$.
$\because BC = AC$,
$\therefore BC^2 = (x+n)(x+m)$.
$\because 2BC^2 = AB^2$,
$\therefore 2(x+m)(x+n) = (m+x+n)^2$.

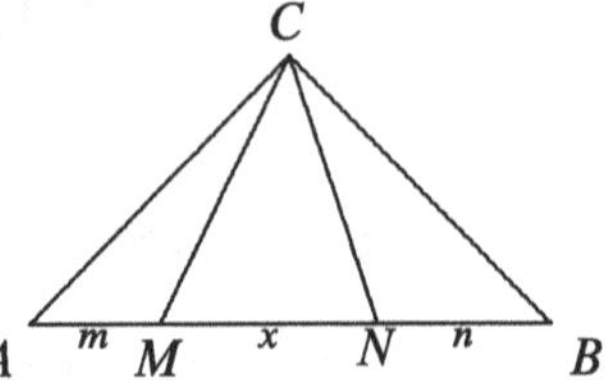

By simplification, we have $x^2 = m^2 + n^2$, therefore the triangle is a right triangle, the answer is (B).

5. Through D introduce $DF \parallel BE$, intersecting AC at F.

$\because AE = 2EC$,
by the midpoint theorem,
$EF = FC$.
$\because AC = 3EC \Longrightarrow AE = 2EC$,
$\therefore AE = 4EF$.
$\because \triangle ADF \sim \triangle AGE$,
$\therefore \dfrac{AG}{GD} = \dfrac{AE}{EF} = 4$.

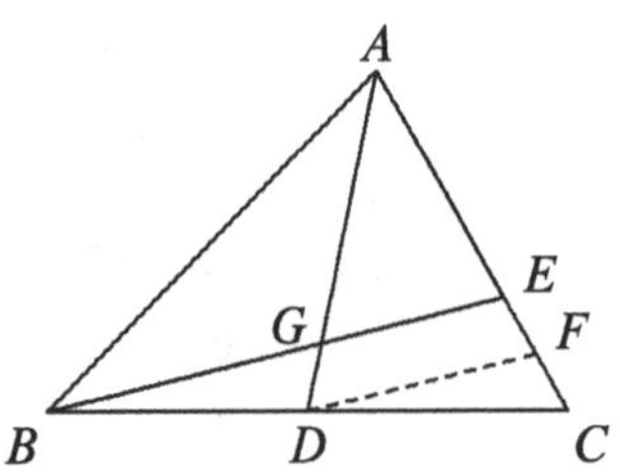

6. From D introduce $DG \parallel BA$, intersecting CF at G. By the midpoint theorem, $CG = GF$ and $DG = \dfrac{1}{2}BF$.

$\because \triangle AEF \sim \triangle DEG$ (A.A.A.),
$\therefore \dfrac{AF}{DG} = \dfrac{AE}{ED} = \dfrac{1}{2}$,
$\therefore AF = \frac{1}{2}DG = \frac{1}{4}BF$,
$\because AF = 1.2$ cm,
$\therefore AB = 4AF + AF = 5AF = 6$ cm.

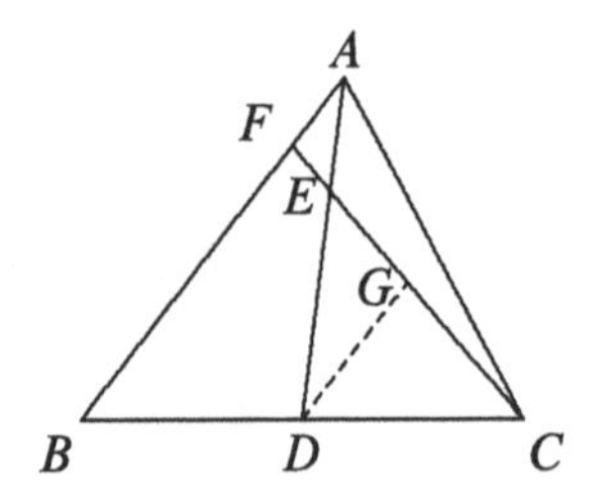

7. Extend DP to meet the extension of CB at F. $\because \triangle PAD \sim \triangle PBF$ (A.A.A.), $BF = \dfrac{PB}{AP} \cdot AD = \dfrac{1}{2} \cdot 2 = 1$, therefore $CF = CB + BF = 3$, $DF = \sqrt{DC^2 + CF^2} = \sqrt{4^2 + 3^2} = 5$.

$$\because DE = \frac{DC^2}{DF} = \frac{16}{5},$$
$$\therefore CE^2 = CD^2 - DE^2$$
$$= 16 - \frac{256}{25} = \frac{144}{25},$$
$$\therefore CE = \frac{12}{5}.$$

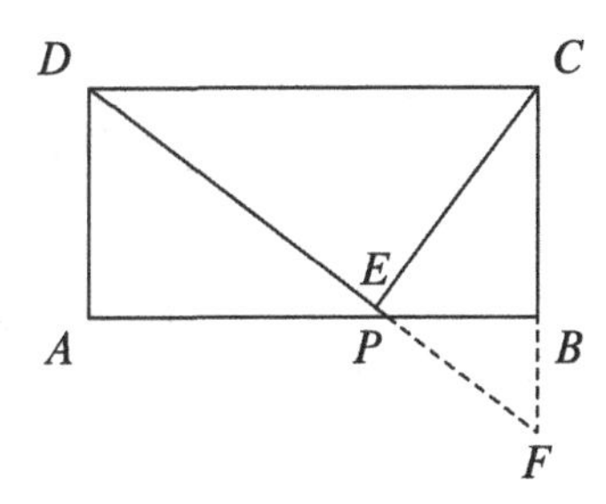

8. $\because AD \parallel BC \Longrightarrow \angle AFE = \angle HAF$, it suffices to show $\angle ACE = \angle FAE$, and for this we show that $\triangle ACE \sim \triangle FAE$ below.

$$\because AE = \sqrt{2}a = \sqrt{2} \cdot EF \text{ and}$$
$$CE = 2a = \sqrt{2} \cdot EA,$$
Besides, $\angle AEC = \angle FEA$,
$\therefore \triangle ACE \sim \triangle FAE$. (S.A.S.).

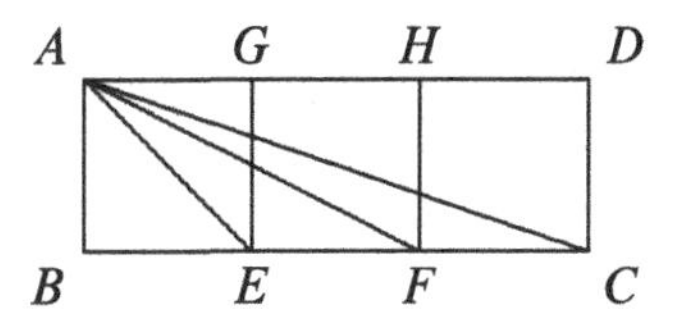

9. From A introduce $AM \perp BC$ at M. Then $Rt\triangle ADM \sim \text{Rt}\triangle CDH$, so

$$\frac{AD}{CD} = \frac{DM}{DH}.$$
$\because \triangle ABC$ is equilateral,
$$\therefore BD + DM = BM = CM = \frac{3}{2}BD,$$
$\therefore DM = \frac{1}{2}BD$, so that
$$\frac{AD}{BD} = \frac{AD}{2MD} = \frac{CD}{2HD} = \frac{BD}{HD},$$
$\therefore \triangle ADB \sim \triangle BDH$ (S.A.S.).
Thus, $\angle DBH = \angle DAB$.

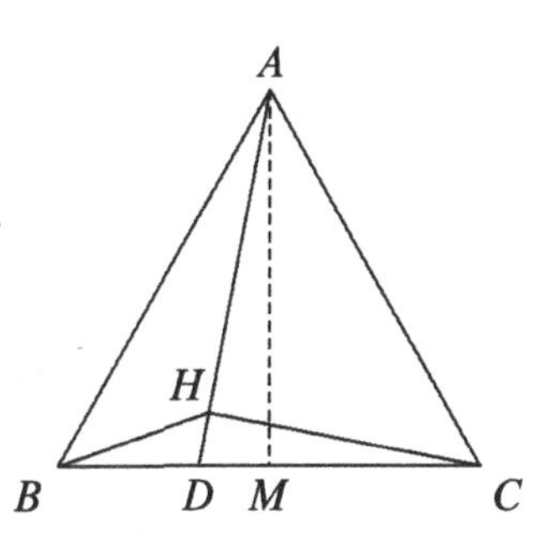

10. Suppose that the angle bisector of $\angle A$ meets BC at D. Since $\angle DAC = \frac{1}{2}\angle A = \angle B$, $\triangle CAD \sim \triangle CBA$ (A.A.A.).

$$\therefore \frac{AC}{BC} = \frac{CD}{AC} \quad \text{or} \quad AC^2 = BC \cdot CD.$$

Since AD bisects $\angle A$ yields $\dfrac{AB}{AC} = \dfrac{BD}{CD}$,

$$\therefore AC^2 + AB \cdot AC$$
$$= AC^2\left(1 + \frac{AB}{AC}\right)$$
$$= BC \cdot CD\left(1 + \frac{BD}{CD}\right)$$
$$= BC(CD + BD) = BC^2.$$

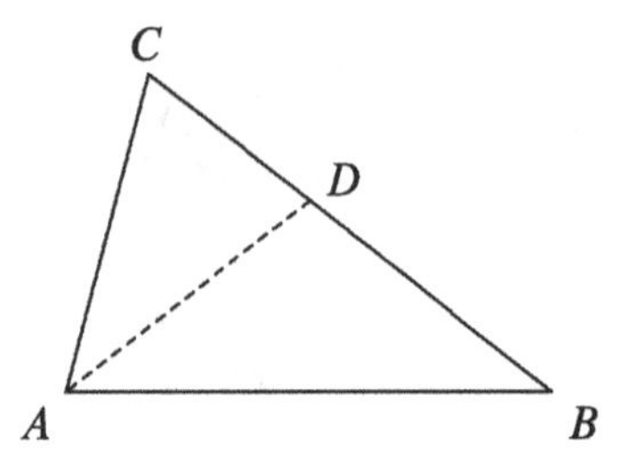

Thus, $AC^2 + AB \cdot AC = BC^2$.

Testing Questions (13-B)

1. Since the three triangles t_1, t_2, t_3 are similar, so the ratios of their corresponding sides are given by

$$PF : DE : PI$$
$$= \sqrt{4} : \sqrt{9} : \sqrt{49}$$
$$= 2 : 3 : 7.$$
$$\therefore CE : DE : BD = 2 : 3 : 7,$$
$$\therefore CE : CB = 2 : (2 + 3 + 7)$$
$$= 2 : 12,$$
$$\therefore [GPF] : [ABC] = 2^2 : 12^2$$
$$= 4 : 144.$$
$$\because [GPF] = 4, \therefore [ABC] = 144.$$

2. $ABCD$ is a rhombus implies that $\angle EAD = \angle DCF = \angle ABC = 60°$. $\because AB \parallel CD, \angle AED = \angle CDF$, therefore $\triangle ADE \sim \triangle CFD$.

$$\therefore \frac{AE}{AD} = \frac{CD}{CF},$$

It follows that $\dfrac{AE}{AC} = \dfrac{AC}{CF}$.

$$\because \angle EAC = \angle ACF = 120°,$$

$$\therefore \triangle EAC \sim \triangle ACF \text{ (S.A.S.)},$$

$$\therefore \angle FAC = \angle CEA.$$

Since $\angle ACE$ is shared by triangles EAC and AMC, $\triangle CEA \sim \triangle CAM$, therefore $\dfrac{CA}{CE} = \dfrac{CM}{CA}$, namely $CA^2 = CE \cdot CM$.

3. From D introduce $DG \parallel BE$, intersecting AC at G, then

$$\frac{EG}{GC} = \frac{BD}{CD} = \frac{2}{3},$$

$$\therefore \frac{AF}{FD} = \frac{AE}{EG} = \frac{AE}{EC} \cdot \frac{EC}{EG} = \frac{3}{4} \cdot \frac{5}{2} = \frac{15}{8}.$$

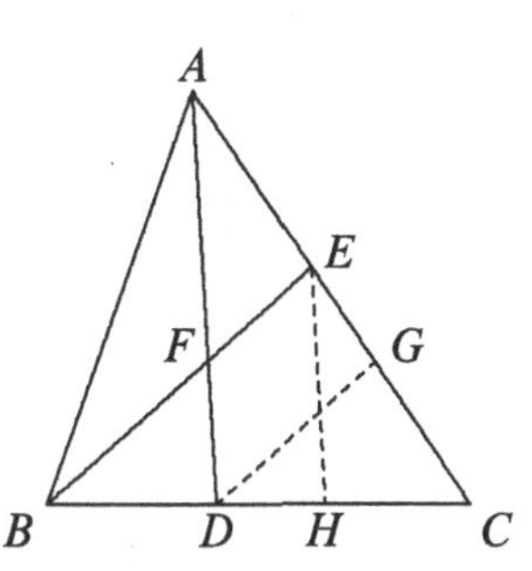

From E introduce $EH \parallel AD$, intersecting BC at H, then

$$\frac{DH}{HC} = \frac{3}{4}, \text{ therefore } \frac{BF}{FE} = \frac{BD}{CD} \cdot \frac{CD}{DH} = \frac{2}{3} \cdot \frac{7}{3} = \frac{14}{9}, \text{ thus}$$

$$\frac{AF}{FD} \cdot \frac{BF}{FE} = \frac{15}{8} \cdot \frac{14}{9} = \frac{35}{12}.$$

4. We define the angles 1 to 6 as shown in the diagram below. Then

$\angle 1 = \angle 2 = \angle 3$, $GO = GB$ and $\angle 4 + \angle 1 = 90° = \angle 2 + \angle 6$, so $\angle 6 = \angle 4 = \angle 5$ which implies $CE = CO$. Since $\triangle COG \sim \triangle FOC$ and $FG \parallel AB$, we have

$$\frac{AF}{CF} = \frac{BG}{CG} = \frac{GO}{CG} = \frac{CO}{CF} = \frac{CE}{CF},$$

hence $CE = AF$.

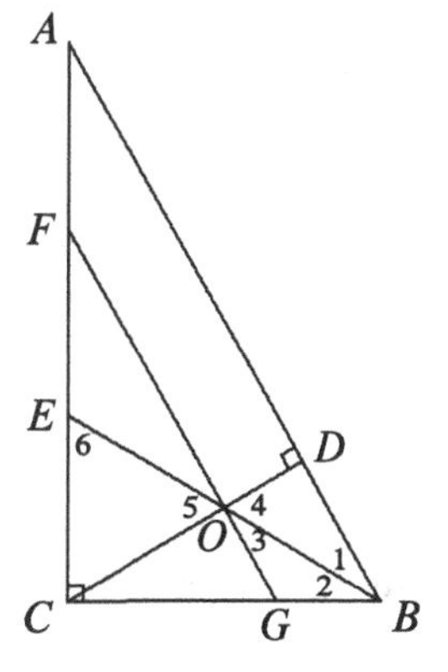

5. Since $BD \parallel MN$, $\triangle DOC \sim \triangle NPC$, $\triangle BOC \sim \triangle RPC$, $\triangle ABO \sim \triangle AMP$, $\triangle ADO \sim \triangle ASP$. therefore we have

$$\frac{PN}{OD} = \frac{CP}{CO} = \frac{PR}{OB}, \quad \frac{PM}{OB} = \frac{AP}{AO} = \frac{PS}{DO}.$$

Therefore we have

$$\frac{PN}{PR} = \frac{OD}{OB}, \frac{PM}{PS} = \frac{OB}{OD}.$$

$$\therefore \frac{PN}{PR} \cdot \frac{PM}{PS} = \frac{OD}{OB} \cdot \frac{OB}{OD},$$

$$\frac{PN \cdot PM}{PR \cdot PS} = 1, \text{ i.e.,}$$

$$PM \cdot PN = PR \cdot PS.$$

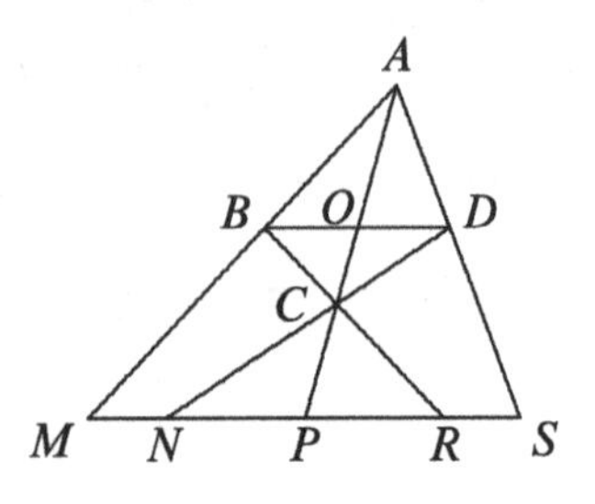

Solutions To Testing Questions 14

Testing Questions (14-A)

1. Connect $B'C$. From $AB = BB'$, $BC' = 3BC$, $[BB'C'] = \frac{BB' \cdot BC'}{AB \cdot BC} \cdot [ABC] = 3$. Similarly,

$$\frac{[AA'B']}{[ABC]} = \frac{AA' \cdot AB'}{AC \cdot AB} = 6,$$

$$\therefore [AA'B'] = 6.$$

$$\frac{[CC'A']}{[ABC]} = \frac{CA' \cdot CC'}{AC \cdot BC} = 8,$$

$\therefore [CC'A'] = 8$. Thus,

$$[A'B'C'] = 3 + 6 + 8 + 1 = 18.$$

2. Connect PO. Since PE, PF are the heights of $\triangle DPO$ and $\triangle APO$, we can use the area method for getting the sum of the two heights. Since $DO = AO$,

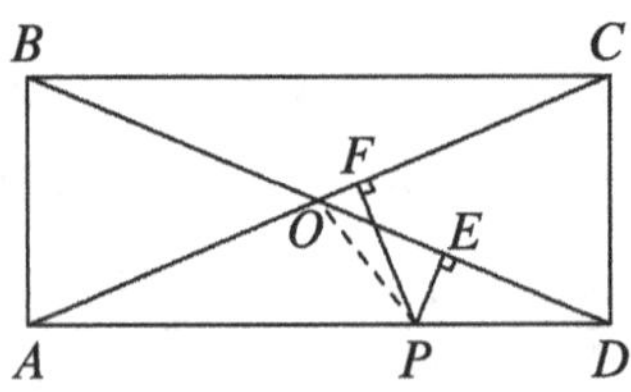

$$(PE + PF) \cdot AO = PE \cdot DO + PF \cdot AO = 2([DPO] + [APO])$$
$$= \frac{1}{2}[ABCD] = \frac{12 \cdot 5}{2} = 30.$$
$$\because AO = \frac{1}{2}\sqrt{5^2 + 12^2} = 6.5,$$
$$\therefore PE + PF = 30 \div 6.5 = \frac{60}{13}.$$

3. Connect PA, PB, PC. Let the length of the side of $\triangle ABC$ be a, then

$$[ABC] = [PAB] + [PBC] - [PAC]$$
$$= \tfrac{1}{2}(h_1 a + h_3 a - h_2 a) = 3a.$$
$$\because [ABC] = \frac{\sqrt{3}}{4}a^2, \therefore a = 4\sqrt{3},$$
$$\therefore [ABC] = 12\sqrt{3}.$$

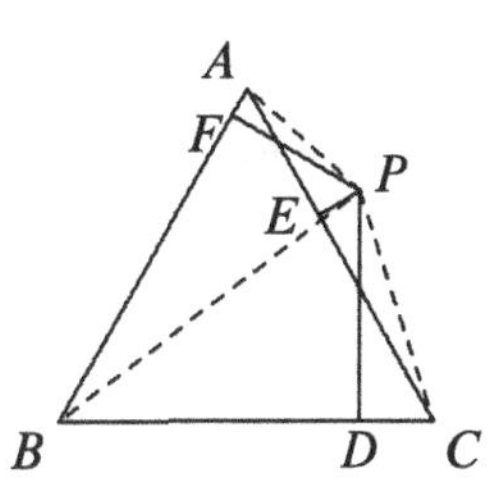

4. Let A be area of the $\triangle ABC$. Then $A = \frac{1}{2}h_a a = \frac{1}{2}h_b b = \frac{1}{2}h_c c$, therefore

$$a = \frac{2A}{h_a}, \qquad b = \frac{2A}{h_b}, \qquad c = \frac{2A}{h_c}.$$

Since $2b = a + c$, we have $\frac{4A}{h_b} = \frac{2A}{h_a} + \frac{2A}{h_c}$, i.e. $\frac{2}{h_b} = \frac{1}{h_a} + \frac{1}{h_c}$.

5. As shown in the digram below, we denote areas of the corresponding triangles by $S_1, S_2, S_3, S_4, S_5, S_6$ respectively. Then $BD = 2DC \Longrightarrow S_3 = 2S_2 = 8$.

$$\because \frac{BG}{GE} = \frac{S_2 + S_3}{S_1} = 4 = \frac{S_4 + S_5}{S_6},$$
$$\therefore S_4 + S_5 = 4S_6.$$
$$\because S_4 + S_5 = 2(S_6 + S_1) = 2S_6 + 6,$$
$$4S_6 = 2S_6 + 6 \Longrightarrow S_6 = 3.$$
$$\therefore S_4 + S_5 = 12.$$
$$\because \frac{S_4}{S_5} = \frac{BF}{FA} = \frac{S_2 + S_3}{S_1 + S_6}$$
$$= \frac{12}{6} = 2,$$

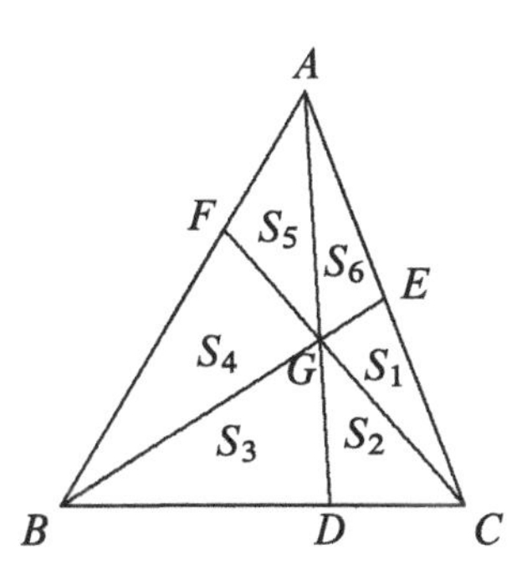

we have $S_4 = 2S_5$, so $S_4 = 8$, $S_5 = 4$. Thus,

$$[ABC] = 3 + 4 + 8 + 8 + 4 + 3 = 30.$$

6. Let $BC = a, CA = b, AB = c$. By the theorem on angle bisector, $\frac{AE}{EC} = \frac{c}{a}$, therefore $AE = \frac{bc}{a+c}$, $EC = \frac{ab}{a+c}$. Similarly,

$$AF = \frac{bc}{a+b}, \quad BF = \frac{ac}{a+b},$$

$$BD = \frac{ac}{b+c}, \quad CD = \frac{ab}{b+c}.$$

$$\therefore \frac{[AFE]}{[ABC]} = \frac{AF \cdot AE}{AB \cdot AC}$$

$$= \frac{bc}{(a+b)(a+c)}.$$ Similarly,

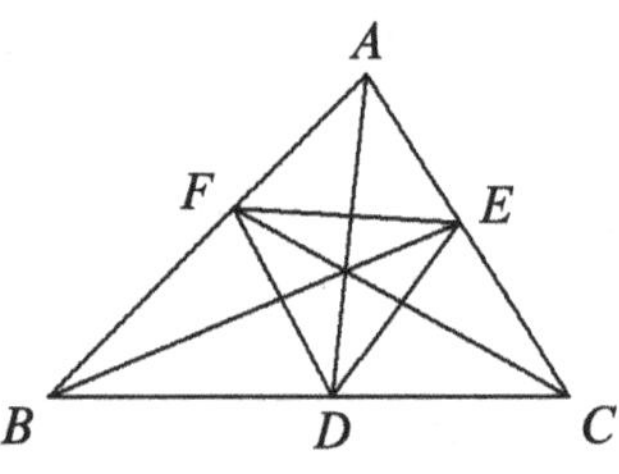

$$\frac{[BDF]}{[ABC]} = \frac{ac}{(b+a)(b+c)}, \quad \frac{[CED]}{[ABC]} = \frac{ab}{(c+a)(c+b)},$$ so.

$$\frac{[DEF]}{[ABC]} = 1 - \frac{bc}{(a+b)(a+c)} - \frac{ca}{(b+a)(b+c)} - \frac{ab}{(c+a)(c+b)}$$

$$= \frac{(a+b)(b+c)(c+a) - bc(b+c) - ca(c+a) - ab(a+b)}{(a+b)(b+c)(c+a)}$$

$$= \frac{2abc}{(a+b)(b+c)(c+a)}.$$

7. Connect AG, FD. Since $AD \parallel BC$, we have $[ABD] = [ACD]$,

$$\therefore [EBD] = [EAD] + [ABD]$$

$$= [EAD] + [ACD] = [EAC].$$

On the other hand, Since $EF \parallel BD$,

$$[EBD] = [FBD] = [ACG],$$

$$\therefore [EAC] = [ACG], \quad \therefore EG \parallel AC.$$

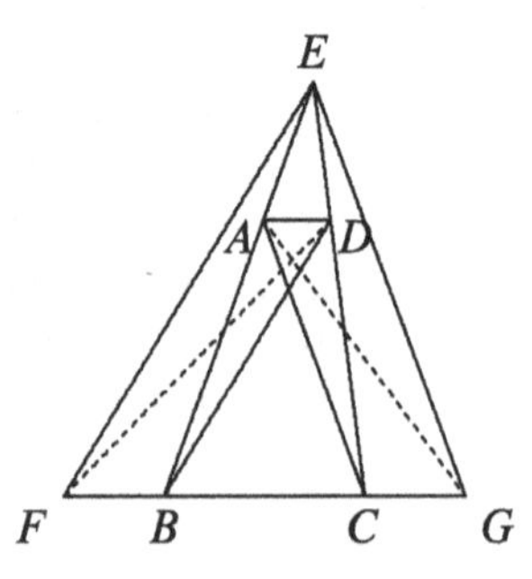

8. Let t_a, t_b, t_c be the perpendicular distance of P from BC, CA, AB, and h_a, h_b, h_c the heights on BC, CA, AB, respectively.

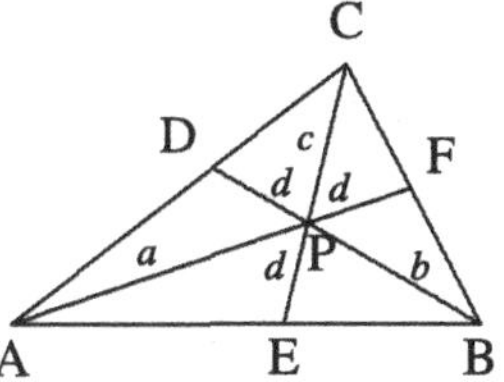

$$\because \frac{t_a}{h_a} + \frac{t_b}{h_b} + \frac{t_c}{h_c} = 1 \text{ and}$$

$$\frac{t_a}{h_a} = \frac{[CPF]}{[CAF]} = \frac{d}{d+a},$$

$$\frac{t_b}{h_b} = \frac{d}{d+b}, \quad \frac{t_c}{h_c} = \frac{d}{d+c},$$

$$\therefore \frac{d}{d+a} + \frac{d}{d+b} + \frac{d}{d+c} = 1,$$

$$d[(b+d)(c+d)+(a+d)(c+d)+(a+d)(b+d)] = (a+d)(b+d)(c+d),$$

$$3[(ab+bc+ca)+6(a+b+c)+27]$$
$$= abc + 3(ab+bc+ca) + 9(a+b+c) + 27,$$
$$\therefore abc = 9(a+b+c) + 54 = 9 \times 43 + 54 = 441.$$

9. From C introduce $CD \perp AC$, intersecting the extension of EF at D.

$$\because \angle ABE = \angle CED,$$
$$\therefore \text{Rt}\triangle ABE \sim \text{Rt}\triangle CED,$$
$$\therefore \frac{[CED]}{[ABE]} = \left(\frac{CE}{AB}\right)^2 = \frac{1}{4}$$
and $\dfrac{CE}{CD} = \dfrac{AB}{AE} = 2.$

Since $\angle ECF = 45^\circ = \angle DCF$, CF is the angle bisector of $\angle DCE$,

therefore the distance from F to CE is equal to that of F to CD, hence $\dfrac{[CEF]}{[CDF]} = \dfrac{CE}{CD} = 2$. Thus,

$$[CEF] = \frac{2}{3}[CED] = \frac{2}{3} \cdot \frac{1}{4}[ABE] = \frac{2}{3} \cdot \frac{1}{4} \cdot \frac{1}{4} = \frac{1}{24}.$$

Testing Questions (14-B)

1. Let CF, BF intersect DE, AE at P, Q respectively. It suffices to show that $S_4 = S_6 + S_2$. Let h_1, h_2, h_3 be the heights of the triangles ABE, FBC, and DEC respectively, then $h_2 = \frac{1}{2}(h_1 + h_3)$. Therefore

$$
\begin{aligned}
S_4 + S_5 + S_1 &= \frac{1}{2}h_2 \cdot BC \\
&= \frac{1}{4}(h_1 + h_3) \cdot BC \\
&= \frac{1}{4}h_1(2BE) + \frac{1}{4}h_3(2EC) \\
&= (S_6 + S_5) + (S_2 + S_1) \\
&= S_6 + S_2 + S_5 + S_1,
\end{aligned}
$$

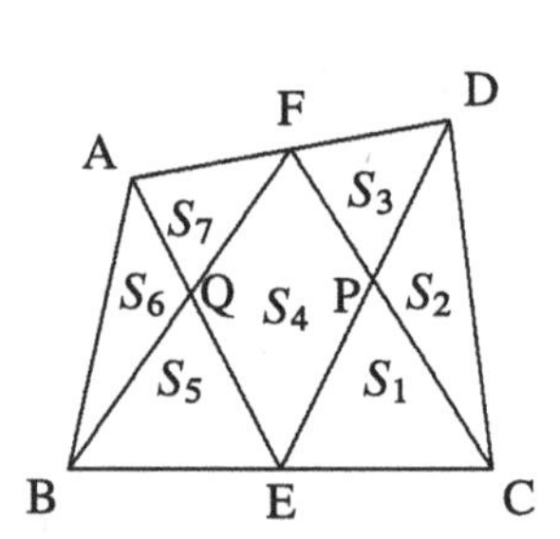

thus, we have $S_4 = S_6 + S_2$.

2. Extend AG to P such that $AG = GP$. Let AP and BC intersect at D, then D is the midpoint of BC, and $GD = DP = \frac{1}{2}AG$. Therefore $BGCP$ is a parallelogram, $BP = GC = 2$.

$$
\begin{aligned}
&\because GB^2 + BP^2 = (2\sqrt{2})^2 + 2^2 \\
&= 12 = GP^2, \quad \angle GBP = 90°, \\
&\therefore BGCP \text{ is a rectangle}, \\
&\therefore [BGC] = \tfrac{1}{2}[BGCP] \\
&= \tfrac{1}{2} \cdot 2 \cdot 2\sqrt{2} = 2\sqrt{2}, \\
&\therefore [ABC] = 3[BGC] = 6\sqrt{2}.
\end{aligned}
$$

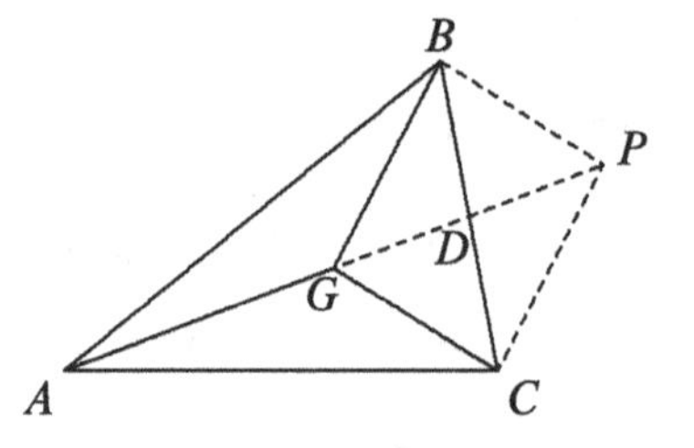

3. Since $\dfrac{BD}{DC} = \dfrac{[ABD]}{[ACD]} = \dfrac{[PBD]}{[PCD]} = \dfrac{[ABD]-[PBD]}{[ACD]-[PCD]} = \dfrac{[APB]}{[CPA]}$, and similarly,

$$
\begin{aligned}
&\frac{CE}{EA} = \frac{[BPC]}{[APB]}, \quad \frac{AF}{FB} = \frac{[CPA]}{[BPC]}, \\
&\therefore \frac{BD}{DC} \cdot \frac{CE}{EA} \cdot \frac{AF}{FB} \\
&= \frac{[APB]}{[CPA]} \cdot \frac{[BPC]}{[APB]} \cdot \frac{[CPA]}{[BPC]} \\
&= 1.
\end{aligned}
$$

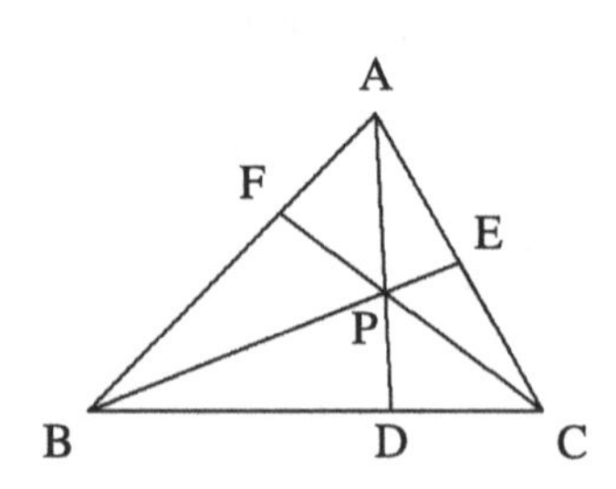

Note: When P, the point of intersection of three lines, is outside the triangle ABC, the conclusion is still true, and it can be proven similarly.

4. Let $x = [BOC], y = [COA], z = [AOB]$. Since $\triangle AOC$ and $\triangle A'OC$ have equal altitudes and $\triangle AOB$ and $\triangle A'OB$ so are also,

$$\frac{AO}{OA'} = \frac{[AOC]}{[A'OC]} = \frac{[AOB]}{[A'OB]}$$
$$= \frac{[AOC] + AOB]}{[A'OC] + [A'OB]} = \frac{y+z}{x},$$

similarly,

$$\frac{BO}{OB'} = \frac{z+x}{y}, \frac{CO}{OC'} = \frac{x+y}{z}.$$

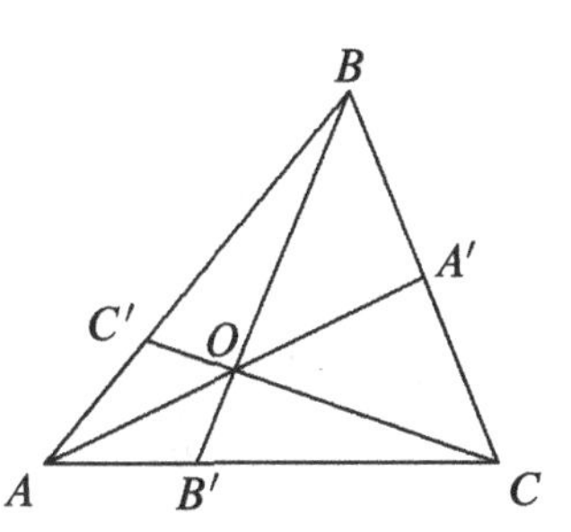

thus,

$$\frac{AO}{OA'} \cdot \frac{BO}{OB'} \cdot \frac{CO}{OC'} = \frac{(x+y)(y+z)(z+x)}{xyz}$$
$$= \frac{yz^2 + +y^2z + x^2z + xz^2 + xy^2 + yx^2 + 2xyz}{xyz}$$
$$= 2 + \frac{y+z}{x} + \frac{x+z}{y} + \frac{x+y}{z}$$
$$= \frac{AO}{OA'} + \frac{BO}{OB'} + \frac{CO}{OC'} + 2 = 92 + 2 = 94.$$

5. From D introduce $DL \parallel AC$, intersecting PB at L. $\because AP = PD$ and $\triangle APE \sim \triangle DPL$, $\triangle APE \cong \triangle DPL$.

$\therefore PL = PE = 3, \quad BL = LE = 6.$
$\therefore D$ is the midpoint of BC.
From D introduce $DK \parallel AB$, where K is on PC, then $\triangle PDK \cong \triangle PAF$,
$\therefore PF = \frac{1}{4}CF = 5, \ CP = 15.$
By the formula for median,
$BC^2 + 4PD^2 = 2(PC^2 + PB^2),$

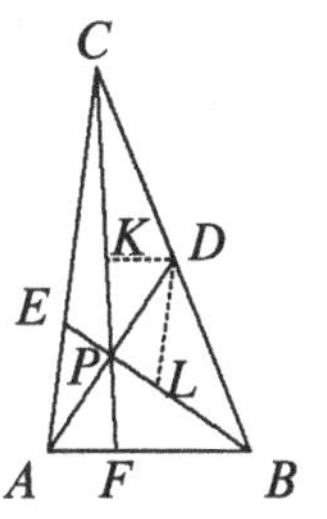

$BC^2 = 2(15^2 + 9^2) - 12^2 = 468$, i.e. $BD^2 = 117 = 9^2 + 6^2 = PB^2 + PD^2$, therefore $PD \perp PB$ at P. Hence $[BPD] = \frac{1}{2} \cdot 6 \cdot 9 = 27.$

Based on the area of $\triangle BPD$ we can get $[ABC]$ as follows:

$[APC] = [CPD] = [BPD] = [BPA] = 27, \quad \therefore [ABC] = 4 \cdot 27 = 108.$

Solutions To Testing Questions 15

Testing Questions (15-A)

1. The two divisions are as follows:

$$\begin{array}{r|l} & 3x^2+6x+7 \\ \hline x-2 & 3x^3-5x+6 \\ & 3x^3-6x^2 \\ \hline & \quad 6x^2-5x+6 \\ & \quad 6x^2-12x \\ \hline & \qquad 7x+6 \\ & \qquad 7x-14 \\ \hline & \qquad\quad 20 \end{array}$$

$$\begin{array}{r|rrrr} 2 & 3 & 0 & -5 & 6 \\ & & 6 & 12 & 14 \\ \hline & 3 & 6 & 7 & \boxed{20} \end{array}$$

The quotient is $3x^3+6x+7$, and the remainder is 20.

2. when using synthetic division to calculate

$$(-6x^4-7x^2+8x+9)\div(2x-1),$$

we carry out $(-6x^4-7x^2+8x+9)\div(x-\frac{1}{2})$ first. Then

$$\begin{array}{r|rrrrr} \frac{1}{2} & -6 & 0 & -7 & 8 & 9 \\ & & -3 & -\frac{3}{2} & -\frac{17}{4} & \frac{15}{8} \\ \hline & -6 & -3 & -\frac{17}{2} & \frac{15}{4} & \boxed{\frac{87}{8}} \end{array}$$

Therefore

$$q(x)=\frac{-6x^3-3x^2-\frac{17}{2}x+\frac{15}{4}}{2}=-3x^3-\frac{3}{2}x^2-\frac{17}{4}x+\frac{15}{8}, r=\frac{87}{8}.$$

3. By the factor theorem, $0=f(-3)=81-81+72+3k+11$, therefore $k=-\dfrac{83}{3}$.

4. Since $f(-1) = f(-2) = 0$, we have $a - b = 3$ and $2a - b = 9$. By solving them we have $a = 6, b = 3$.

5. From the remainder theorem, we have $f(x) = (x-1)q_1(x) + 1$.
   Let $q_1(x) = (x-2)q_2(x) + r_1$, then
$$f(x) = (x-1)(x-2)q_2(x) + r_1(x-1) + 1.$$
   Since $f(2) = 2$, we have $r_1 + 1 = 2$, i.e. $r_1 = 1$, hence
$$f(x) = (x-1)(x-2)q_2(x) + x.$$
   Let $q_2(x) = (x-3)q_3(x) + r_2$, then
$$f(x) = (x-1)(x-2)(x-3)q_3(x) + r_2(x-1)(x-2) + x.$$
   Since $f(3) = 3$, we have $2r_2 + 3 = 3$, i.e. $r_2 = 0$. Thus,
$$f(x) = (x-1)(x-2)(x-3)q_3(x) + x,$$
   the remainder of $f(x)$ is x when divided by $(x-1)(x-2)(x-3)$.

6. Let $x^5 - 5qx + 4r = (x-2)^2(x^3 + ax^2 + bx + c)$, then
   $x^5 - 5qx + 4r = (x^2 - 4x + 4)(x^3 + ax^2 + bx + c)$
   $= x^5 + (a-4)x^4 + (4+b-4a)x^3 + (4a+c-4b)x^2 + (4b-4c)x + 4c,$
   therefore
$$a-4 = 0,\ \ 4+b-4a = 0,\ \ 4a+c-4b = 0,\ \ 4b-4c = -5q,\ \ 4c = 4r.$$
   From them we have orderly $a = 4, b = 12, c = 32, q = 16$ and $r = c = 32$. Thus, $q = 16,\ r = 32$.

7. Let $f(x) = ax^3 + bx^2 + cx + d$. From assumptions,
$$\begin{aligned} f(x) &= (x^2-1)q_1(x) + 2x - 5 \ \text{ and} \\ f(x) &= (x^2-4)q_2(x) - 3x + 4. \end{aligned}$$
   Let $x = \pm 1$, it follows that
$$\begin{aligned} a+b+c+d &= -3, & (15.43) \\ -a+b-c+d &= -7. & (15.44) \end{aligned}$$
   Let $x = \pm 2$, it follows that
$$\begin{aligned} 8a+4b+2c+d &= -2, & (15.45) \\ -8a+4b-2c+d &= 10. & (15.46) \end{aligned}$$

By (15.43) + (15.44) and (15.45) + (15.46), respectively, we obtain

$$b + d = -5, \quad (15.47)$$
$$4b + d = 4. \quad (15.48)$$

Therefore, (15.48) − (15.47) yields $b = 3$ and $d = -8$. By substituting back the values of b and d into (15.43) and (15.45), respectively, we obtain

$$a + c = 2, \quad (15.49)$$
$$4a + c = -3. \quad (15.50)$$

(15.50) − (15.49) yields $a = -\frac{5}{3}$, and from (15.49), $c = \frac{11}{3}$. Thus,

$$f(x) = -\frac{5}{3}x^3 + 3x^2 + \frac{11}{3}x - 8.$$

8. Let $f(x) = x^3+7x^2+14x+8$. Since all the coefficients are positive integers, $f(x) = 0$ has negative roots. 8 has negative divisors $-1, -2, -4, -8$, and it is clear that $f(-1) = 0$, so -1 is a root, i.e. $(x + 1)$ is a factor. By synthetic division, we obtain

$$f(x) = (x + 1)(x^2 + 6x + 8).$$

It is easy to see that $x^2 + 6x + 8 = (x + 2)(x + 4)$, so

$$x^3 + 7x^2 + 14x + 8 = (x + 1)(x + 2)(x + 4).$$

9. The given expression is symmetric in x and y, so it can be expressed in the basic symmetric expressions $u = x + y$ and $v = xy$. Therefore

$$\begin{aligned} &x^4 + y^4 + (x + y)^4 \\ &= (x^2 + y^2)^2 - 2x^2y^2 + (x + y)^4 = (u^2 - 2v)^2 - 2v^2 + u^4 \\ &= 2u^4 - 4u^2v + 2v^2 = 2(u^4 - 2u^2v + v^2) = 2(u^2 - v)^2 \\ &= 2((x + y)^2 - xy)^2 = 2(x^2 + y^2 + xy)^2. \end{aligned}$$

10. The given expression is a cyclic polynomial. Define $f(x) = xy(x^2 - y^2) + yz(y^2 - z^2) + zx(z^2 - x^2)$, where y, z are considered as constants, then

$$f(y) = yz(y^2 - z^2) + zy(z^2 - y^2) = 0,$$

so $(x - y), (y - z), (z - x)$ are three factors of the given expression. Since the given polynomial is homogeneous and has degree 4, the fourth factor is linear homogeneous cyclic expression, so must be $A(x + y + z)$. Hence

$$xy(x^2-y^2)+yz(y^2-z^2)+zx(z^2-x^2) = A(x+y+z)(x-y)(y-z)(z-x).$$

Let $x = 2, y = 1z = 0$, then $6 = -6A$ i.e. $A = -1$. Thus,

$$xy(x^2-y^2)+yz(y^2-z^2)+zx(z^2-x^2) = (x+y+z)(x-y)(y-z)(x-z).$$

Testing Questions (15-B)

1. From f is a common factor of g and h, f is a common factor of $3g(x) - h(x) = 4x^2 - 12x + 4 = 4(x^2 - 3x + 1)$, so, by the factor theorem,

$$4(x^2 - 3x + 1) = A(x^2 + ax + b),$$

where A is a constant. By the comparison of the coefficient of x^2, $A = 4$. Thus $a = -3, b = 1$, and $f(x) = x^2 - 3x + 1$.

2. For $f(y) = y^m - 1$, since $f(1) = 0$, so $f(y)$ has factor $y - 1$, i.e. $y^m - 1 = (y - 1)q(y)$. Let $y = x^3$, we have

$$x^{3m} - 1 = (x^3)^m - 1 = (x^3 - 1)q(x^3) = (x - 1)(x^2 + x + 1)q(x^3),$$

i.e. $x^2 + x + 1$ is a factor of $x^{3m} - 1$. Therefore $x^{3n+1} - x = x(x^{3n} - 1)$ and $x^{3p+2} - x^2 = x^2(x^{3p} - 1)$ both have the factor $x^2 + x + 1$ also. Thus,

$$x^{3m}+x^{3n+1}+x^{3p+2} = (x^{3m}-1)+(x^{3n+1}-x)+(x^{3p+2}-x^2)+(x^2+x+1)$$

has the factor $x^2 + x + 1$.

3. From the given conditions we have $f(a) = a$, $f(b) = b$, $f(c) = c$. Let $r(x)$ be the remainder of $f(x) - x$ when divided by $(x - a)(x - b)(x - c)$. If $r(x)$ is the zero polynomial, the conclusion is proven.

Below we prove by contradiction that $r(x)$ is the zero polynomial. Otherwise, then its degree is not greater than 2, and

$$f(x) - x = (x - a)(x - b)(x - c)q(x) + r(x),$$

so $r(a) = r(b) = r(c) = 0$. Therefore, the polynomial $r(x)$ has at least three distinct real roots a, b, c, although its degree is not greater than 2. Thus, $r(x)$ is equal to 0 identically, a contradiction.

4. Let the given expression be $P(x, y, z)$. Then P is cyclic. Consider it as a polynomial $f(x)$ of x only and let $x = y$, then

$$f(y) = (y^2 - z^2)(1 + y^2)(1 + yz) + (z^2 - y^2)(1 + yz)(1 + y^2) = 0,$$

so $(x-y)$, and hence $(x-y)(y-z)(z-x)$ are factors of P. The remaining factor is a cyclic polynomial of degree three (but it is non-homogeneous). So

$$\begin{aligned} P(x,y,z) &= (x-y)(y-z)(z-x)[A(x^3+y^3+z^3) \\ &\quad +B(x^2y+y^2z+z^2x)+C(xy^2+yz^2+zx^2) \\ &\quad +Dxyz+E(x^2+y^2+z^2)+F(xy+yz+zx) \\ &\quad +G(x+y+z)+H], \end{aligned}$$

where A, B, C, D, E, F, G, H are the coefficients to be determined. Since the highest index of each of x, y, z on the left hand side is 3, so in the brackets the power of x, y, z cannot be greater than 1, hence $A = B = C = E = 0$.

The comparison of coefficients of x^2y indicates that $H = 0$;

The comparison of coefficients of xy^3 indicates that $G = 1$;

The comparison of coefficients of x^3y^2 indicates that $F = 0$.

Therefore the right hand side is only $(x-y)(y-z)(z-x)(x+y+z+Dxyz)$. Letting $x = 3, y = 2, z = 1$, then

$$-24 = -2(6+6D) \Longrightarrow D = 1.$$

Thus, the factorization of the given expression is

$$(x-y)(y-z)(z-x)(x+y+z+xyz).$$

5. The given conditions gives that

$$f(x) = x^3+2x^2+3x+2 = g(x)\cdot h(x)+h(x) = h(x)[g(x)+1].$$

It is easy to find that $f(-1) = 0$, so $f(x)$ has the factor $x+1$. By synthetic division, we obtain

$$\begin{aligned} &x^3+2x^2+3x+2 \\ &= (x+1)(x^2+x+2) = (x+1)[(x^2+x+1)+1] \\ &= (x^2+x+1)(x+1)+(x+1). \end{aligned}$$

Since h is not a constant, and its degree is less than that of g, so it must be a linear polynomial, and g is a quadratic polynomial with integer coefficients. Thus,

$$g(x) = x^2+x+1, \quad h(x) = x+1$$

satisfy all the requirements. Since the coefficient of x^3 is 1, and all the coefficients of g are integers, the solution is unique.

Index